Anonymous

Report on the Scientific Study of the Mental and Physical Conditions of Childhood

Anonymous

Report on the Scientific Study of the Mental and Physical Conditions of Childhood

ISBN/EAN: 9783744693578

Printed in Europe, USA, Canada, Australia, Japan

Cover: Foto ©berggeist007 / pixelio.de

More available books at **www.hansebooks.com**

REPORT

ON THE

SCIENTIFIC STUDY

OF THE

MENTAL AND PHYSICAL CONDITIONS OF CHILDHOOD.

WITH PARTICULAR REFERENCE TO CHILDREN OF DEFECTIVE CONSTITUTION; AND WITH RECOMMENDATIONS AS TO EDUCATION AND TRAINING.

(The Report is based upon the Examination of 50,000 Children seen in 1888-91, and of another 50,000 seen in 1892-94.)

LONDON:

PUBLISHED BY THE COMMITTEE, PARKES MUSEUM, MARGARET STREET, W.

1895.

MEMBERS OF THE COMMITTEE ON THE MENTAL, AND PHYSICAL CONDITION OF CHILDREN.

President.

LORD EGERTON OF TATTON, Late Chairman of the Royal Commission on the Blind, Dumb, &c.

Chairman.

Sir DOUGLAS GALTON, K.C.B., D.C.L., LL.D., F.R.S.

Treasurer.

[1] ROWLAND HAMILTON, F.S.S.

[1] H. ASHBY, M.D., F.R.C.P.
[1] FLETCHER BEACH, M.B., F.R.C.P.
Rev. GEO. BELL, Head Master Marlborough College.
G. W. BLOXAM, M.A.
[2] E. W. BRABROOK, F.S.A.
[1] T. BRIDGWATER, M.B., LL.D.
[6] Dr. BURGERSTEIN, Vienna.
Surgeon-General CORNISH, C.I.E., F.R.C.S.
[5] Sir THOS. CRAWFORD, K.C.B., Q.H.S., M.D., LL.D. (deceased).
JOSEPH R. DIGGLE, M.A., J.P.
[1] J. L. H. LANGDON DOWN, M.D., F.R.C.P.
R. FARQUHARSON, M.D., LL.D., M.P.
J. G. GARSON, M.D.
[3] TIMOTHY HOLMES, M.A., F.R.C.S.
[1] W. W. IRELAND, M.D.
[6] Dr. JACOBI, New York.
[1] HOLMAN, C., M.D., M.R.C.S.
[1] Sir GEO. HUMPHRY, M.D., F.R.S.
[6] Dr. KOTELMAN, Hamburg.
[6] Dr. KUBORN, Seraing-Liège.
Sir PHILIP MAGNUS, B.Sc., B.A.
General MOBERLY.
SHIRLEY MURPHY, M.R.C.S., M.O.H., London County Council.

[1] F. NEEDHAM, M.D., M.R.C.P., Commissioner in Lunacy.
J. W. PALMER.
Miss POOLE.
G. V. POORE, M.D., F.R.C.P.
Sir RICHARD QUAIN, Bart., F.R.S.
HENRY RAYNER, M.D., M.R.C.P.
T. L. ROGERS, M.D., M.R.C.P.
[6] Dr. ROUSSEL, Paris.
[1] G. H. SAVAGE, M.D., F.R.C.P.
J. HOLT SCHOOLING, F.S.S.
Rev. T. W. SHARPE, C.B., Sen. Chief Inspector, Education Department.
[1] G. E. SHUTTLEWORTH, B.A., M.D., M.R.C.S.
W. WILBERFORCE SMITH, M.D., M.R.C.P.
T. A. SPALDING, LL.D.
Hon. E. LYULPH STANLEY, M.A.
[5] J. C. STEELE, M.D. (deceased).
Dr. OCTAVIUS STURGES (deceased).
[1] D. H. TUKE, M.D. (deceased).
CARDINAL VAUGHAN, Archbishop of Westminster.
[2] E. WHITE WALLIS, F.S.S.
[1][2] FRANCIS WARNER, M.D., F.R.C.P.
Rev. J. C. WELLDON, Head Master of Harrow School.
[1] D. YELLOWLEES, M.D., LL.D.

Hon. Secretary.

E. WHITE WALLIS, F.S.S.

Offices.

PARKES MUSEUM, Margaret Street, London, W.

The following bodies are represented on the Committee:—

1 The British Medical Association.
2 British Association.
3 The Charity Organisation Society.
4 The Royal Statistical Society.
5 The Sanitary Institute.
6 Foreign Representatives.

EXECUTIVE COMMITTEE.

Sir Douglas Galton, K.C.B., F.R.S., *Chairman*.

Fletcher Beach, M.B., F.R.C.P.

T. Bridgewater, M.B., LL.D.

J. Langdon Down, M.D., F.R.C.P.

Rowland Hamilton, F.S.S.

C. Holman, M.D., M.R.C.S.

J. W. Palmer.

Miss Poole.

G. H. Savage, M.D., F.R.C.P.

J. Holt Schooling, F.S.S.

G. E. Shuttleworth, B.A., M.D., M.R.C.S.

F. Warner, M.D., F.R.C.P.

E. White Wallis, F.S.S., *Secretary*.

STATISTICAL COMMITTEE.

J. Holt Schooling, F.S.S.

J. W. Palmer.

J. Langdon Down, M.D., F.R.C.P.

Francis Warner, M.D., F.R.C.P.

E. White Wallis, F.S.S.

PREFACE.

The object of the Committee in undertaking this investigation of the mental and physical condition of childhood was to furnish a reliable statement of existing conditions found among the pupils attending public elementary and other schools, and thus to establish a scientific basis for the study of the requirements of child-life, providing also information for the guidance of the State, educational authorities, and philanthropic bodies.

The scientific field of study opened up in the Report is in many respects entirely new; the Committee have endeavoured so to arrange their voluminous records that they may be available for the purposes of research in many important directions.

By the co-operation of the Charity Organisation Society, and British Medical Association, it has been possible to include in the present report certain statistics relative to 50,000 children seen during the earlier enquiry, in addition to the 50,000 seen under the auspices of this Committee, thereby affording in main principles a wider basis of evidence. The Committee also recognise their indebtedness to the British Medical Association for continued money grants, and to The Sanitary Institute for the use of offices. Their thanks are also due to the British Association for money grants, and to the London School Board for facilities afforded them for the examination of children.

Among the efforts made to secure public attention to this important subject may be mentioned—Letters by Sir Douglas Galton and other members of the Committee, published in the *Times ;* Papers read by Dr. Francis Warner before the Royal Statistical Society, the British Association and other societies ; and a meeting held in June 1894 at the residence of Lord Egerton of Tatton, the President of the Committee, when proceedings were inaugurated for approaching the House of Lords, with the object of obtaining an official enquiry into the conditions of child-life.

In placing this report, the result of much arduous work, before the public, the Committee hope that their efforts may induce the State authorities to establish some permanent means of ensuring that the several matters therein alluded to may receive due attention, in the interests alike of the child-population, and of the community at large.

LIST OF CONTENTS.

VII

TABLES.

PAGE

VIII

SCIENTIFIC STUDY

OF THE

MENTAL AND PHYSICAL CONDITIONS OF CHILDHOOD.

CHAPTER I.

In 1888, at the Annual Meeting of the British Medical Association in Glasgow, a Committee was appointed to conduct an investigation as to the average development and condition of brain power among the children in primary schools. The first action taken was to decide upon a mode of procedure. A form of schedule was drawn up for use in reporting, and a paper of "suggestions" was prepared by Dr. Warner and others, containing a list of points worthy of note in making observations.

In 1889, a report was published on 5,444 children seen in fourteen schools, as drawn up by Dr. Francis Warner, who for over ten years had previously studied the scientific methods of observing and describing indications of mental status in children.

The Council of the British Medical Association have warmly supported the investigation and continues to do so, and have made grants of money each year.

In the same year, 1889, evidence was given by members of the Committee before the Royal Commission on the blind, dumb, etc., and for the first time it was officially recognised that there was a class of children who, while not imbecile, present a certain amount of mental deficiency.

The Commission reported:—

"That with regard to 'feeble-minded' children, they should be separated from ordinary scholars in public elementary schools, in order that they may receive special instruction, and that the attention of school authorities be particularly directed towards this object."

In 1890, a Committee was appointed by the Charity Organisation Society, including members of the former Committee—"To promote a

A

scientific inquiry with regard to the number and condition of feeble-minded children or adults."

The Committee writes as follows:—

"There is a large class of persons which the word 'feeble-minded' may well describe, who are often in distress and often lapse into destitution and degradation, and for whom at present it is usually extremely difficult to make any charitable or other proper provision. They cannot be termed 'imbecile,' nor can they be dealt with under the Idiots Act of 1886. Rather are they in childhood the 'backward' or 'mentally dull' pupils of the school, whose backwardness or dullness is the result of physical causes; and later in life they become, if they are fortunate, the dependents of kind-hearted people, or more frequently the habitual inmates of workhouses, from which they go out from time to time, often to their lasting harm and mischief, and in the case of girls and young women not infrequently to their disgrace and ruin."

"The question of trying to prevent this 'feeble-mindedness' by a better care of children, and of making for feeble-minded and desti-tute adults some provision that may at least preserve them from mischief, has been considered by several societies. The Metropolitan Association for Befriending Young Servants has frequently had to cope with the difficulty in cases of district school and other girls under their charge. The National Vigilance Society has through its Preventive Committee made inquiry of the number of 'feeble-minded' girls and women in workhouses and infirmaries. Their returns, which, though incomplete, are useful, show that they received replies from 203 Boards of Guardians; that during the year 1889, 715 ' weak-minded' women passed through 105 workhouses, and that at 56 workhouses it was stated that the approximate number of such women who were leading immoral lives was 366. The Reformatory and Refuge Union has had the question brought specially before it by resolutions passed at the recent Conference of Managers of Reformatory and Industrial Schools at Glasgow; and the Council of the Charity Organisation Society has also more than once in the course of the past year or two considered the difficulty of dealing with cases of this kind, as well as with cases of epileptics, crippled and deformed persons."

The report of this Committee showed (1) the large number of children below the average physical and mental development; (2) the importance of grouping main classes of defects for the purposes of research ; (3) the importance of ascertaining the co-relations of main classes of defectiveness, and also the co-relations of the individual defects; (4) the greater proportion of children with mental defect among those with several conditions of physical defect; (5) the unequal distribution of defectiveness among groups of schools, the high proportion among the pauper class, also differences in the nationalities.

An interim report was presented by the Charity Organisation Society to the Congress of Hygiene and Demography, 1891, and their full report has since been published.*

Papers were read at the Congress on the work done up to that date, and the present Committee was formed under a resolution passed at the meeting.

In 1892 a full report on 50,000 children, seen in 106 schools by Dr. Francis Warner, in conjunction with other medical members of the Committee, was presented to the Local Government Board.

Since 1892 the work has been in the hands of the present Committee, and another 50,000 children have been examined for this Committee by Dr. Francis Warner, in conjunction with Dr. Shuttleworth, Dr. Fletcher Beach, and others; the methods of observation have been essentially the same as those used since 1888, and their value has been amply proven by experience. The methods of compilation of facts have been much improved, and are fully explained in the report.

The Milroy Lectures of the Royal College of Physicians were delivered by Dr. Francis Warner in 1892, upon the results of enquiry in 106 schools, seen in 1888-91.

In the year 1892 a Committee was appointed by the British Association for the Advancement of Science, for similar enquiries, and the members have worked with the present Committee; three short reports have been issued by the Committee of the British Association, and it has each year been re-appointed.

At the meeting of the International Congress of Hygiene and Demography at Budapest in 1894, Dr. Francis Warner and Dr. Shuttleworth attended as delegates, and presented a short report on the work done; papers on the same subject were read before the Sections, and a demonstration of the methods of examination was given before the Congress, some children having been assembled for the purpose.

---

* "The Feeble-minded, Epileptic, Deformed, and Crippled," and "The Feeble Child and Adult;" and "Report on Feeble-minded, Epileptic, and Cripples," 1892. C. O. S. Series. 15 Buckingham, Street, Strand, London, W.C.

In November, 1894, evidence was given by Dr. Francis Warner before a Departmental Committee on Poor Law Schools, showing in detail the conditions of the Poor Law children in contrast with those in day schools, and making certain recommendations.

Articles have been contributed to journals; press correspondence and notices have from time to time appeared, and although the Committee have hitherto devoted very little attention to this branch of the work, it is evident that the investigation is favourably viewed by the various sections of the public interested in the welfare of the children.

ORGANISATION AND WORK OF THE COMMITTEE.

The Committee now consists of representatives in Hamburg, Paris, New York, Vienna, Seraing-Liege, with representatives from the British Medical Association, the Charity Organisation Society, The Sanitary Institute, and the Royal Statistical Society, together with a number of professional men skilled in education and in the compilation of statistical facts.

Committees appointed by the British Medical Association and by the British Association for the Advancement of Science have joined this Committee, and both these Associations have contributed towards its funds.

The Executive Committee and Sub-Committees were appointed to deal with certain details of the work, and they have supervised the enquiry in all its stages. The Committee has endeavoured to spread a knowledge of the objects and work, but have been hampered for want of necessary funds. Applications have been made to the Local Government Board, the London County Council, most of the City Companies and other public bodies, and also to the Prussian Government, for grants in aid of the work of the Committee, but their applications have been so far without result.

The Statistical Sub-Committee have held a number of meetings to discuss and direct the modes of arrangement and tabulation, and have carefully examined and checked the statistical work and tables.

Following a suggestion made by the Committee to the Education Department, the Rev. T. W. Sharpe, Senior Chief Inspector of the Department, was, at their request, present at the inspection of some of the Schools, and expressed his favourable opinion as to the usefulness of the investigation, and a wish that teachers might learn something of the methods of observing children for the purpose of classifying them.

The Committee supported a memorial of the Council of the British Medical Association concerning the appointment of a Commission for the scientific inquiry as to the condition of the school-population, which was forwarded to the Education Department and Local Government Board, and although so far without result, will be again pressed as opportunity may occur.

Information concerning Irish children as seen in England was forwarded to Dublin Castle at the request of the Inspectors in Lunacy.

Statistics concerning feeble-minded children have been supplied at the request of the London School Board, who have experienced difficulties in connection with such children. A report on 35,361 children seen in 35 Board Schools, by permission of the Board, has been forwarded to the Board.

A letter was addressed to the Universities, Colleges, and other Educational Bodies, suggesting the desirability of establishing lectures on the study of children. One County Council has offered to meet half the expenses of such lectures as a branch of Technical Education, and it is hoped that the London County Council may do the same.

In December, 1893, the Committee issued an Interim Report on the work being done, copies of which were circulated.

A very valuable addendum to vital statistics might be obtained by following up the history of certain cases recorded, by subsequent periodical inspections, but as this is beyond the power of the present Committee, it can only be suggested as one among many other directions in which enquiry may be pushed in the hands of official Commissions.

LIST OF SCHOOLS EXAMINED.

The 106 schools examined in the first enquiry, 1888-91, containing 50,027 children are divided into the following ten Divisions.

DIVISION 1—*Poor Law Schools.*

This includes 19 Poor Law (district and separate) schools, being nearly all of those situated within the London district.

They are under the medical and educational supervision of the Local Government Board. The Roman Catholic children were mostly Irish.

Numbers seen . . . . . .   Boys 5884.   Girls 3947.
Numbers noted . . . .   ,,   1332.   ,,   685.

DIVISION 2— *Certified Industrial Schools.*

This includes 9 schools, they mostly receive children sent by the magistrates as having some connection with crime, either through their own acts, or those of their parents or others.

These schools were mostly near London, they are under the supervision of the Government Home Office.

Some of these were Roman Catholic schools containing mostly Irish children.

Numbers seen .... Boys 1588. Girls 407.
Numbers noted ...... „ 500. „ 91.

DIVISION 3— *Homes and Orphanages.*

This includes 6 schools or philanthropic institutions for the boarding, clothing, and training of orphan children.

They are voluntary and independent of the Government.

Numbers seen ...... Boys 774. Girls 1049.
Numbers noted ...... „ 172. „ 186.

DIVISION 4— *All Resident Schools.*

This includes the 34 schools in the preceding groups combined. In these schools the children receive board, clothing and education.

They are institutions, some under the Government, others voluntary for the care of children more or less destitute.

Numbers seen ...... Boys 8246. Girls 5403.
Numbers noted .... „ 1994. „ 962.

DIVISION 5— *Public Elementary (and other) Day Schools.*

This includes 72 day schools almost all public elementary schools; a few of rather higher grade are included, 10 were under the London School Board, most were voluntary.

Numbers seen.... Boys 18,638. Girls 17,740.
Numbers noted .. „ 3,575. „ 2,645.

DIVISION 6—20 *Day Schools of the Upper Social Class.*

Numbers seen ...... Boys 5281. Girls 4934.
Numbers noted...... „ 1122. „ 796.

DIVISION 7—52 *Day Schools of Poorer Social Class.*

Numbers seen.... Boys 13,357. Girls 12,806.
Numbers noted .. „ 2,435. „ 1,849.

N.B.—Groups 6 and 7 combined make up Group 5.

DIVISION 8—*English Day Schools.*

This includes all day schools containing mostly English children, not Irish or Jews; resident schools with English children are not included in this group.

Numbers seen.... Boys 16,932. Girls 15,875.
Numbers noted .. ,, 3,252. ,, 2,379.

DIVISION 9—*Schools containing Irish Children.*

This includes 3 Poor Law resident schools, 4 certificated industrial resident schools, and 1 public elementary day school.

Numbers seen........ Boys 1694. Girls 595.
Numbers noted ...... ,, 585. ,, 115.

DIVISION 10—*Jewish Day Schools.*
The Jewish Schools, Whitechapel.

Mostly very poor, and children of foreign immigrants; a portion only of the children was seen.

Numbers seen ...... Boys 1389. Girls 1572.
Numbers noted ...... ,, 247. ,, 218.

N.B.—A further analysis of the children of these Nationalities is given in Tables XIX., XX.

The distribution of the children presenting the "defects" described in the nomenclature (see page 72) among the Divisions of schools is given in Table XIII., and the distribution of "Groups of children" described in the catalogue (see page 82), is presented in Table XIV.

In the enquiry made by this Committee 1892-94, 63 schools were examined containing 50,000 children; these schools are arranged in 12 Divisions.

DIVISION I.—7 *London Board Schools, Upper Social Class,*
English Children.

These schools represent English children, not Irish or Jews.

Numbers seen ...... Boys 4800. Girls 4316.
Numbers noted ... ,, 838. ,, 679.

DIVISION II.—11 *London Board Schools, Average Social Class,*
English.

The children in these schools were not characterised as distinctly belonging to either upper or poorer social class. They represent English children.

Numbers seen ...... Boys 6113. Girls 5628.
Numbers noted .... „ 1159. „ 944.

DIVISION III.—14 *London Board Schools, Poorer Social Class, English.*

The children in these schools were of poorer families than those in the preceding group, they consisted mainly of English children.

Numbers seen ...... Boys 6342. Girls 5213.
Numbers noted .... „ 1155. „ 863.

DIVISION IV.—3 *London Board Schools, Jewish children of the Poorer Social Class.*

In these schools almost all the children were Jews.

Numbers seen ...... Boys 1368. Girls 1581.
Numbers noted .... „ 249. „ 223.

DIVISION V.—2 *Country Board Schools, English.*

Board Schools of Seaford, Sussex, and at Harrow. English country children of average social class.

Numbers seen ...... Boys 528. Girls 482.
Numbers noted .... „ 105. „ 70.

DIVISION VI.—1 *Edinburgh Board School, Scotland.*

Scottish children of Upper Social Class.

Numbers seen ...... Boys 803. Girls 807.
Numbers noted .... „ 163. „ 128.

DIVISION VII.—4 *Voluntary Schools, Upper Social Class, English.*

Two schools were in the country, one was a high-class boarding school for boys, and the other two were elementary schools in town.

Numbers seen .... Boys 1232. Girls 939.
Numbers noted ...... „ 250. „ 162.

DIVISION VIII.—5 *Voluntary Schools, Average Social Class, English.*

All were London elementary schools.

Numbers seen .... Boys 968. Girls 988.
Numbers noted ...... „ 228. „ 158.

DIVISION IX.—2 *Voluntary Schools, Poorer Social Class, English.*

National schools in London.

Numbers seen .... Boys 697. Girls 720.
Numbers noted ...... „ 137. „ 96.

DIVISION X.—10 *Voluntary Schools, Irish Children, Poorer Social Class.*

Irish children seen in London.

    Numbers seen ......   Boys 2171.   Girls 1952.
    Numbers noted ....   ,,   535.   ,,   324.

DIVISION XI.—3 *Voluntary Schools, Jewish Children, Average Social Class.*

London Jewish children.

    Numbers seen ....   Boys 823.   Girls 693.
    Numbers noted ......   ,,   138.   ,,   104.

DIVISION XII.—1 *Voluntary School, Jewish Children, Poorer Social Class.*

The Jews' Free school, Whitechapel. This is the same school as was examined in 1891, but care was taken that different children were examined, and only part of the school was seen.

    Numbers seen ....   Boys 440.   Girls 394.
    Numbers noted ......   ,,   155.   ,,   78.

These twelve divisions of schools are arranged so as to be comparable with those seen in 1888-91, and the terms used in describing them bear the same significance as those previously used. Summation of the groups here given will afford information concerning all the board schools and voluntary schools, as well as of those of the social classes and of nationalities.

. Such information may be needed in future enquiries, and many additional divisions of schools may be prepared by summation of those given.

CHAPTER II.

EXPLANATION OF THE METHODS USED IN THIS ENQUIRY, AND COMMENTS ON THE REPORT.

Methods of Observation and Research.

The methods of examining the physical condition of children seen in schools must necessarily be more limited than those used in the consulting room. Arrangements already exist in the reports of H. M. Inspectors for determining the intellectual acquirements of school children; their family history and evidence as to their home life could not be obtained, and school managers naturally object to questions being asked of the children concerning their health, which, if put, would not be likely to elicit any trustworthy information; it is also impracticable to handle the children for the purposes of physical examination.

The observer must therefore depend mainly upon inspection, and having determined beforehand what points to look to, he must record accurately what he sees.

The importance of deciding on a number of physical signs for observation and record was appreciated before the work was commenced on a large scale.

For the purpose of observing those finer balances and reactions of the nerve system which indicate neuro-mental potentialities, it is better to deal with the children in a uniform manner, and not to handle them.

The terms used in giving descriptions of children should each indicate a fact seen and capable of verification and comparison, the essence of scientific description.

Viewing a child, we see its body and some of the indications of brain action.

The signs observed are of two kinds: (A) points of form, proportion and indications of type of development of the body and its separate parts ; thus, looking at the child, we note the cranium and the separate features, the ears, the nose, the palpebral fissures and the mouth, *i.e.*, its physiognomy; also its growth and the indications of nutrition.

(B) Nerve signs are seen in the balance of the head, the spine, the upper extremity and the digits, as well as in the facial action and eye movements. These movements and balances of action or postures are observed as signs of the action and condition of the nerve-centres. Inspection of the children is most conveniently conducted in a large and well-lighted room, the pupils being drawn up in ranks, a standard at a time, or in groups of about forty, so that the inspector may view each individual. It is convenient to fix the child's eyes by asking each in turn to look at an object held up, *e.g.*, a shilling at the end of a pencil. The trained observer can read off the physiognomy of the individual features and their parts, the facial condition and eye movements, the balance of the head and body, &c., as quickly as a printed line.

The children in the group are then requested to hold their hands out straight, the action being shown them momentarily ; the balance is noted as a further indication of the nerve-condition. Finally, the palate is inspected in every child.

At each stage of the inspection, children presenting deviations from the normal are asked to stand aside ; any dull children or special cases not picked out by inspection may now be presented by the teacher ; the selected cases are kept and the other children are dismissed to the class-room.

Each of the selected cases is then reviewed individually and described on a schedule form ; the teachers' report concerning the child is taken down or filled in by them afterwards. Occasionally some detailed enquiry concerning a child may be made, or some brief mental examination may be desirable, but as a rule no questions are asked of the child.

This method works smoothly and uniformly with a minimum of trouble to the teachers and pupils. The teachers almost universally acknowledged that the dull children had been selected by inspection, and very few were subsequently presented by them.

In taking notes of the cases selected by inspection as presenting some points of deviation from the normal, a schedule form was filled in for every child.

On the schedule form the first column is headed "Development, Physiognomy, &c.," and as sub-headings the words Palate, Ears, Growth, are printed; these are ticked if the parts of child corresponding are found normal on inspection, and abnormalities are described. The second column is headed "Movements, Postures, &c.," with sub-headings "Expression, Orbicularis oculi, Eye movements, Hand balance, Head balance, General balance." In a column headed "Physical Health and Nutrition," particulars of this class are recorded.

The teachers give their opinion of the child as to mental status in the column "School Report." The more important points of the case are entered by the inspector in the final column. Such schedule was filled in for each of the children noted as presenting some deviations from the normal. At the same time the name, age, and standard in school is entered by an assistant on the name-sheet, with a reference number, which is repeated on the schedule, and thus the case is identified in subsequent analysis. The number of children seen in each standard (boys and girls) and the date of the visit, together with the names of any visitors present, are endorsed on the school report.

For the purpose of preparing statistics, each case as described in the written report taken in the school, is entered in a Register in which the headings of the columns indicate the defective conditions observed, "Cranial Abnormality," "Palate Defective," "Expression Defective," "Low Nutrition," etc; the case being entered under such headings as correspond to its defects. Further columns were added to the Register as additional defects were observed in the progress of the enquiry. Thus the headings of the columns form the list of defective conditions given in the *Nomenclature* (see page 72).

There are four main classes or divisions into which the defective conditions observed may be grouped.

A. Defects in Development of the body and its parts; in size, form, or proportioning of parts. (See Nomenclature of defects, 1 to 42.)

B. Abnormal Nerve-signs; certain abnormal actions, movements, and balances. (See Nomenclature of defects, 43 to 63.)

C. Low Nutrition, as indicated by the child being thin, pale or delicate.

D. Mental Dulness. The teachers' report as to mental ability was added to the record of each child registered, and those stated to be below the average in ability for school work were registered as "Dull."

The importance of these Main Classes of defect is indicated in the following Table.

TABLE I. *(cases seen 1888—91).—Showing Number of Children with each Main Class of defect, and the percentage of each class respectively that presented conditions of defect co-related thereto. Percentages are taken upon the numbers given. For numbers of cases presenting the combined conditions, see Catalogue on page 82. For facts corresponding in cases seen 1892—94, see Table* XXIV.

Number presenting main Class of Defect alone or in combination.	With Group 28, Defect in Development alone or in combination.		With Group 29, Nerve Signs alone or in combination.		With Group 30, Low Nutrition alone or in combination.		With Grp. 31. Mental Dulness alone or in combination.	
Boys. Girls.	Boys.	Girls.	Boys.	Girls.	Boys.	Girls.	Boys.	Girls.
Defect in Development, Group 28. 3616 2235	...	...	54·6	49·0	20·2	32·0	38·3	41·5
Nerve Defect. Group 29. 3413 2074	57·8	52·8	...	...	18·6	28·8	40·1	42·4
Low Nutrition. Group 30. 1030 973	71·1	74·6	61·0	61·4	...	...	39·0	40 5
Mental Dulness. Group 31. 2216 1463	63·0	63·4	61·8	60·1	18·1	27·0	...	...

Many of the children present two or more defects, and consequently occur more than once in this and similar Tables.

A. Defects in bodily development are frequently found to be co-incident with brain defects, usually lowering mental status. The connecting link between defects of body and defective mental action is the co-incident defect of brain which may be known by observation of "abnormal nerve-signs." It is to this view of the question as demonstrated by the original researches of Dr. Francis Warner, that the Committee attach great weight. It is the coincident observation of Conditions of development of the body, and "nerve-signs" indicating brain-action, that forms a special feature of the present investigation, and distinguishes the methods used from older physiognomical research.

Another fact co-related with defect in development is the tendency of such cases, especially girls, to become pale, thin, and delicate. It is in the co-relation of abnormalities in the proportion of parts of the body, with abnormal nerve-signs, low nutrition and mental dulness, that we find a criterion of the really defective status connected with the abnormality and the value of such signs. This is illustrated as follows:—

TABLE II. (*cases seen* 1888-91).—*Showing Number of Children with Defects in Development in Resident and Day Schools, and the percentage of these Cases that presented co-related conditions taken upon the Cases with Defects in Development.*

Number of Development Cases.	Percentage with Low Nutrition.			Percentage with Abnormal Nerve Signs.		
	Boys.	Girls.	Total.	Boys.	Girls.	Total.
Resident Schools, I.–XXXIV.— B. 1324. G. 671. T. 1995.........	15·7	19·8	16·6	61·9	52·6	58·8
Day Schools, XXXV.–CVI.— B. 2292. G. 1564. T. 3856	23·3	37·8	29·2	50·3	47·5	49·2
Total for all Schools, I.–CVI.— B. 3616. G. 2235. T. 5851	20·2	32·0	24·9	54·6	49·0	52·4

Number of Development Cases.	Percentage with Low Nutrition and Nerve Signs.			Percentage reported as Dull.		
	Boys.	Girls.	Total.	Boys.	Girls.	Total.
Resident Schools, I.–XXXIV.— B. 1324. G. 671. T. 1995.........	9·0	11·3	9·7	39·8	44·2	41·3
Day Schools, XXXV.–CVI.— B. 2292. G. 1564. T. 3856	12·7	19·5	15·5	37·9	40·3	38·7
Total for all Schools, I.–CVI.— B. 3616. G. 2235. T. 5851	11·3	17·0	13·5	38·3	41·5	39·7

In reporting for the purpose of selecting "the worst made children" a definitely high standard must be maintained, the most marked cases only being noted; on the other hand, for the purpose of determining the causation of defects, a lower standard of defectiveness should be adopted, ranging from the minimum that can be distinctly observed and described to the maximum of the pathologist.

If conditions of defective development are looked for in children brought for examination, on account of their known physical or mental feebleness, a very much higher co-relative value will be put upon the individual signs observed than that obtained in this enquiry where each feature was observed in all the school children.

Many cases with one or two defects are otherwise perfectly average children; it is by an association of signs in development and in the nerve system that we recognise the defective condition of the children.

B. Nerve signs are of value as they indicate brain states; their general significance has been dwelt upon.

Many of these signs indicate action and balance of nerve centres at the moment of observation; the same thing may be said of mental tests, and here is presented a body of signs for clinical use, such as represent the finer shades of conditions among the nerve centres.

In all cases described in the notes the nerve-sign was looked for on two occasions; when inspecting the children in rank, and again when describing the individual case; in most cases its repeated occurrence only was recorded, if not repeated the sign was not noted.

The co-relations of abnormal nerve-signs respectively with defects in development, low nutrition, and mental dulness are given; the first and second may be looked upon as possibly having a casual connection with the nerve sign.

C. Nutrition as applied to the children in this enquiry implies that the child as seen was thin, pale, or delicate looking. It is not sufficient evidence as to good nutrition to look at the face only, this part may be well nourished and yet the limbs may be thin; it is usually well to feel the child's arms or legs.

The most important fact noted with regard to these cases of low nutrition, is that 73 per cent. were cases presenting visible signs of deviations from the normal in development of the features and other parts.

It seems then that there is a large group of children, amounting to nearly 3 per cent. of the children seen, who are so defective in make as to be usually of low nutrition when seen in school.

It appears that these children are of lower general constitutional power and tend to an ill-nourished condition under the stress of life and the many causes of mental excitement which, while they render them sharper mentally, militate against nutrition of the body and its tissues.

That the amount of mental stimulus received by children does lower their general nutrition seems to be further indicated as follows :—

On dividing 36,000 day scholars, seen 1888-91, into two groups, 10,200 seen in day schools of upper social class, presumably well fed children, 5·2 per cent. were found to be of low nutrition, and among the 25,000 children in the poorer day schools 3·9 per cent. ; the only explanation offered is that the upper class children have more stress upon them than those of poorer social position.

The statistical methods employed were not the same in the earlier as in the later portions of the enquiry.

During 1888—91, for the purpose of preparing statistics of co-incident or combined conditions, a list of all cases presenting each principal condition was prepared from the register, and parallel columns were headed with the names of such conditions as it was desired to show in co-relation with it. Thus the list of "cranial abnormalities" had columns by its side, headed "abnormal nerve-signs" and " nutri-tion low," etc., in which the case was also entered if these conditions had been registered against the child.

Proceeding systematically in this way, the co-relation of the more important defects was obtained.

Actuarial methods of arranging statistics were employed 1892—94, which, while simpler in form than those used earlier, afford a much more nearly complete and useful analysis of cases.

The Statistical Sub-Committee, taking the Register as the basis of this actuarial work, prepared a card of the form shown on the next page, on which the list of defects is printed, such card was filled in for each case registered, presenting all the information contained in the Register of the case, the defects present being indicated by drawing the pen through the name of the defect. A formulated epitome at the right hand lower corner of the card indicates the main division of defect in the case.

FACSIMILE OF RECORDING CARD.

School...................   *Card No.*............

St^{d.}......................   *Reg. No.* ....   **GIRLS.**

Age...................   .... *Spl. Rep^{t.}*...............

A	DEVELOPMENT DEFECTS	47	O. oculi lax
a 1	CRANIUM	48	Eye movements
2	Large	49	Head balance
3	Small	50	Hand weak
4	Bossed	51	Hand nervous
5	Forehead	52	Finger Twitches
6	Frontal ridge	53	Lordosis
		54	OTHER NERVE-SIGNS
b 11	EXTERNAL EAR		
c 12	EPICANTHIS	**C**	NUTRITION
d 13	PALATE	**D**	DULL
14	Narrow	**E**	EYE-CASES
15	V-shaped	64	Squint
16	Arched	65	Glasses plus
17	Cleft	66	Glasses minus
18	Other types	67	Myopia, no glasses
e 19	NASAL BONES	68	Cornea disease
f 20	GROWTH SMALL	69	Eye, lost accident
g 21	OTHER DEVELMT. DFTS.	70	Eye, lost disease
B	NERVE-SIGNS	**F**	RICKETS
43	General balance	**G**	EXCEPTIONAL CHILDREN
44	Expression	*i* 82	CRIPPLES
45	Frontals overact		
46	Corrugation		**A B C D E F G**

In sorting the cards for the purpose of compiling the Tables relating to the cases seen 1892-94, the defects included under E, F, and G were not considered. Thus a card with A and F marked was counted as A only; B, C, and E as B C only; and so on. Combinations of A, B, C, or D, with E, F, or G, have not been tabulated.

CHAPTER III.

METHODS OF STUDYING THE RESULTS OF OBSERVA-TIONS RECORDED WITH REFERENCE TO TABLES.

Having described and enumerated the cases observed, we proceed to study the records obtained. This may be done by means of Analysis, Grouping, Co-relation, and the Distribution of Conditions.

Analysis.—We may take a case and regard only one of its defects. This is done in the Nomenclature of Defects (see page 72), in which the numbers of children presenting each defect respectively are given independently of the consideration of whatever else the child may present. In the Catalogue of Groups of Cases (see page 82) the Primary Groups 13 to 27 are obtained by analysis, and give the smallest groups of cases, viz., those presenting only one or more main classes of defects. Groups 28 to 55, containing more than one main class of defects, are larger.

Grouping.—Cases are grouped in various ways. This mode of study is explained and illustrated in the catalogue (see page 82). Many other methods of grouping cases might be arranged for special purposes, and the tables are so arranged as to be available for research in any direction which science may suggest. The analysis of all cases into the 16 Primary Groups (see catalogue) will be found convenient in many respects. Analyses of groups of cases may be distributed in various ways, and in the process co-relations may be obtained by taking proportions.

Co-relation.—It is in the co-relations of defects that new informa-tion is mostly to be looked for, supplying evidence as to the real significance of the defects respectively, and as to their causation. Much attention has been given to the methods, and facilities are afforded for further actuarial research in this direction. Inasmuch as it has been shown by the comparison of groups of schools that the co-relation of the main classes of defects varies as to degree, with the character of the environment, it is advisable to determine the per-centage of co-relations of defects upon similar groups of cases under different environment. To some extent this has been done by giving the co-relations of the main classes of defects, and individual defects, as seen in children in day schools and in residential schools. The difference in the numerical values of these co-relations under different environments are in some degree a measure of their effect.

Table XV. gives individual defects as observed with their co-relations, the cases being distributed for divisions of schools seen 1888-91. In the summary of Tables XVI. and XVII. for all schools (I. to CVI.) seen 1888-91 the co-relations are given in percentages.

Tables XI.. XII., and XVIII. give binary combinations of defects. Certain co-relations are also given in The British Medical Association's Report. Other co-relations may be obtained by comparison of the groups of cases.

The co-relation of a main class of defects may be studied by analysis of the class into two groups, in which one contains all cases with the condition to be co-related, the other division containing the remainder of cases without that condition. The percentage of the first division of cases taken upon the number in the main class of defect presents the percentage co-relation. Thus:—To determine the percentage co-relation of Development cases with Nerve-signs.

Group 28, All Development cases, 9,777.

Group 32, All Development cases with Nerve-signs, 4,545.

The co-relation of Development cases with Nerve-signs is the percentage of the latter upon the former number = 46 per cent.

An average co-relation upon 100,000 children can thus be calculated for any group of cases from the figures given in the Catalogue of Groups of Cases.

Distribution of Cases.—It follows from what has been said as to co-relation that it is useful in research to give the distribution of Groups of Cases.

Table IX. gives distribution of the defects in Divisions of Schools for children seen 1888-91.

Table X. gives distribution of the Groups of Cases in Divisions of Schools for children seen 1888-91.

Table XVIII. gives distribution of binary combinations of defective development for Groups of Schools 1888-91.

Tables XIX., XX., give distribution of various groups of cases among the children of the Nationalities.

In all Tables children seen with the enquiry 1888-91 are kept separate from those seen 1892-94, but so far as possible it is arranged that the set of cases may be comparable and capable of summation.

CHAPTER IV.

THE SIGNS OBSERVED IN DESCRIBING CHILDREN.

The principal signs of defect are here described, with remarks as
to their significance. The numbers in brackets refer to the nomenclature given on page 72, in which all defects are defined and enumerated ; their co-relations are given in Tables XV., XVI., XVII.,
XVIII., for the cases seen 1888-91.

(1). *Cranial abnormalities.*—Of all defects in development, abnormalities of the cranium appear to be the most important, being the
most numerous, and having the highest pathological co-relations.
The size and probable volume of the brain is a point of first-class
importance, and the size of the cranium is in children a fair indication
of the size of the brain.

The following standard of the normal in a well-developed child of
good potentiality may be given: Head circumference at nine months,
17·5 inches, at twelve months, 19 inches, at seven years, 20 to 21
inches. This is probably a high standard of the normal, too high if
deviations therefrom are to be considered as pathological; after
three years of age 19 inches cranial circumference is too small ; in
this investigation no head of any age was described as small which
was up to a circumference of 20 inches. In estimating the volume
of a cranium it is preferable to proceed by inspection ; noting its
form and not solely the circumference or other measurements; good
estimate of both form and volume may be obtained by placing the
open hand on the child's head, subsequently using a tape measure.
It also appears that defects in form of the cranium are often core-
lated with brain deficiences. Defects of the cranium may be divided
into sub-classes.

(2). *Large Heads.*—It seems probable that a large proportion of
these cases resulted from rickets at an earlier period, for of 168 cases
of rickets with defect of cranium, 55·3 per cent. were large heads.

(3). *Small Heads.*—In this group, contrary to the usual rule, the
defect is more common among girls. If there be no other defect,

mental faculty may be average, but the child usually remains thin and delicate ; such children may, in after life, undertake good work and do it, but are more liable than others to exhaustion, migraine, and breakdown of the nerve-system. At school these children are often delicate and irregular in attendance from ailments.

(4). *Cranial Bosses.*—Cranial bosses are most usual at the site of the ossific centres of the two halves of the frontal bone; they may occur at the sides of the head over the parietal centres and elsewhere, as well as at the site of the anterior fontanelle. As to the frontal bosses which are the more common, they are not always outgrowths or thickenings of bone ; it is often found on section that they depend upon a thrusting out of the frontal bone, a forward projection of the wall of the cranium independent of bone thickening, and apparently due to abnormal proportion in the bone growth. Occasionally these bosses are unequal in development producing asymmetry of the head. These bosses are probably largely due to rickets ; were all possible means adopted for prevention of rickets, we should probably have fewer children with cranial abnormalities and defects co-related thereto.

(5). *Defect of Forehead.*—All marked defects of the frontal bone, other than bosses or frontal ridge and asymmetry, are here grouped as a rather miscellaneous collection. The most common defect is narrowness, the forehead being often at the same time shallow, but cases of wide and overhanging forehead were included in the same group.

(6). *Frontal Ridge.*—The vertical suture between the two halves of the frontal bone may be the site of a bony ridge present in all degrees up to the prow shaped forehead of the scapho-cephalic child. Apparently in its lesser forms this sign is not very important, unless it be associated with a contracted forehead or other defect of cranium.

(7). *Head Asymmetrical.* As already said this asymmetry may be due to one large lateral cranial boss or angular development; it was much more common to find the forehead on the left side sloping away towards the temple.

The defect does not appear to be of more consequence than other irregularities of the cranium when found in children otherwise normal.

(8). *Dolicho-cephalic.*—These long heads are usually also large, and the condition itself appears to be hardly pathological, certainly many of these children were bright and intelligent.

(10). *Other types of Cranium.*—This group includes—7 cases of hydrocephalus: boys 5, girls 2 ; also oxycephalic, or elevated and conical heads, and others with the anterior portion of the skull much larger than the posterior segment. As these cases did not appear to afford data for precise knowledge without detailed examination, they are here passed over.

(11). *Ears defectively made.*—Deviations from the normal in the size and proportions of the parts of the external ears appear to be defects in development, and not due to mechanical pressure in infancy.

The most usual defect is an ear large and outstanding, coarse in its cutaneous covering, and red or bluish, often with slight varicosities, the antehelix is often absent or imperfectly formed, and the helix may be absent or contracted forming a cave-like ear. Such ears, like cold blue hands, are liable to chilblains. The helix may be absent partially, as in the middle part of the margin or for the whole rim. Such defects are usually symmetrical, but not always.

As to the adherent lobe of the ear it was often seen, but not specially in ears otherwise defective, and no reason appeared for considering it a marked sign of defect. *Supernumerary ears* were represented by outgrowths in front of the tragus, and depressions, apparently representing remnants of branchial clefts, were occasionally seen at the upper termination of the helix. There were two or three children with almost total absence of the concha, which was represented by a cartilaginous growth only. In one girl the ear was displaced several inches by the contraction of a cicatrice.

(13). *Defective condition of Palate.*—Defect of the palate, though less frequent than that of the cranium, stands next to it, as having an almost equally high pathological co-relation. Some facts will be given as to the concurrence of defects of the palate with the cranium and nasal bones as well as with rickets.

The principal defects of palate are in its proportions as seen in the horizontal or vertical plane. Without being otherwise altered the palate may be contracted laterally or narrow. The V shaped palate is pointed more or less sharply at its anterior extremity, the alveolar processes being nearly straight lines meeting anteriorly at an acute angle. The high-arched or vaulted palate deviates from the normal in the vertical plane. No observations are offered as to the relative value of these types. Among less common forms are the flat or low palate, in some cases with forward projection of the upper incisors, which are strongly inclined forwards in equine fashion in place of being vertical; this type is not common.

The suggestion having been made that defects of the nasal bones might be associated with defect of the palate, the following particulars are given from enquiry, 1888-91 :—

	Boys.	Girls.	Total.
Palate defective in cases of rickets	23	8	31
„ and cranium defective in rickets	21	6	27
„ defective, and nasal bones thick, wide, or sunken ...	14	15	29
„ and cranium defective, with nasal bones thick, wide or sunken	4	3	7
Cranium abnormal, and nasal bones thick, wide, or sunken ...	36	29	65

Palate Crypts.—One or more small pits or puncta are sometimes seen on either side the raphe of the hard palate at its posterior extremity, or they may be represented by slender lines sloping outwards and forwards, looking almost like scars. We do not know their origin or significance, but they do not appear to be either a normal development or an effect of disease. They were not looked for in all palates, as a good light and convenient circumstance are needed for their detection.

(19). *Nasal Bones Defective.*—The bony bridge of the nose may be wide and thick, or it may be sunken and depressed. It appears to be sometimes a temporary condition. A family of children was seen in whom this feature was marked in the younger children only and we were told that the older members presented the same condition in their earlier years. This condition has been referred to in describing palates; in only 4 cases of Rickets was this noted.

(20). *Children small for age.*—The co-relations appear to indicate that small grown children are at a disadvantage. Many of the children with small heads were small in growth also; in such cases the child was registered under both headings, but it will be seen that the number of children with small heads was much larger than the number with small growth. This is an example where normal proportion in the body is not to the child's advantage; the small headed child is probably better fitted for after life when its growth is otherwise normal.

Some of the cases appear to be due to Rickets.

Among the smaller groups to be dealt with, the co-relation value varies greatly. It seems as though some of the less common defects were greater indications of departure from the normal, than those more commonly seen.

(21). *Cases presenting "other defects in development."*—This group contains the signs of defective development not printed as headings in the columns of the Register; the group is kept together for the sake of convenience in dealing with "Binary defects in development." Further work in determining the value of these signs respectively is obviously desirable ; some appear to have a high co-relative value.

The number of cases in many of these sub-groups is too small to suggest a complete analysis of them, but the co-relations of the most numerous or the most interesting are given.

Sub-groups are as follows:—

Cases seen 1888—91.

(30). *Hands blue and cold*	.. Boys 17.	Girls 11.	Total 28.
With nerve-signs	„ 14.	„ 8.	„ 22.
With low nutrition	„ 3.	„ 4.	„ 7.
With defect in development	„ 12.	„ 6.	„ 18.
Dull	„ 9.	„ 6.	„ 15.

This group though not numerous demands attention, as those children presented many defects. This condition appears to be more common in resident than in day schools ; it appears independent of weather, though cold may increase it and cause chilblains. The cases were seen in the following months—March 1st, July 2nd, September 2nd, October 16th, November 7th. Among these cases were : Cranial abnormalities 5, Defective palate 3, Icthyosis 1, Ptosis 1, Mentally defective 5.

(39). *Palpebral Fissures, small.*—The eyelids may be small as well as the opening between them, both in the transverse and vertical measurement. In some cases the opening is not symmetrical, being wider on the inner than in the outer half; the transverse axis may slope downwards and outwards, or upwards and outwards in place of being horizontal. In these cases we sometimes find the mouth also small, so that the face being of the normal size, the space between its openings appears large, giving a peculiar aspect or physiognomy.

We now pass on to the abnormal nerve-signs observed.

(44). *Expression defective.*—We may describe the visible muscular action seen in a face, and still there may be an expression in it which entirely baffles description in anatomical terms. Further, a face may be balanced or moved abnormally by the action of certain muscles, and yet it may carry upon it a good expression. We may describe action in the frontal muscles, the corrugators, the orbicularis oculi, &c., and over and above this we have the general expression of the face superadded. Certain terms are useful in describing expression; there may be a fixed expression, want of variation, *i.e.*, one fixed uniform action or balance of muscular tone; or we may have to use more general terms, such as "defective," "bad." There may be no expression, *i.e.*, none other than that indicated by form or modelling of the features.

(45). *Frontals Overacting.*—The frontal muscles almost always act symmetrically, at the same time and in similar degree; their action produces horizontal creases in the forehead, which may be deep if these act strongly. Sometimes the muscles are seen working under the skin in vermicular fashion, with an athetoid movement; in other cases the action is fine, producing minute creases, and what might be called a dull forehead. This over-muscular action does not necessarily erase expression. Such overaction may be seen in children from earliest infancy upwards; the condition may be temporary, and having lasted a sufficient number of years to produce creases in the forehead, it may pass away. These muscles are often more quiet when the child is at work or being talked to than when let out to play; the mental attitude termed quiet attention is that under which the frontal area is the most quiet.

(46). *Corrugation.*—Corrugation, or knitting of the eyebrows, is due to overaction or hypertonicity of the corrugator muscles; vertical creases are produced by overaction, and a fine wrinkling of the skin producing local dulness is sometimes seen. This sign seems more closely associated than any other single sign with some forms of mental stress, and may be seen in children suffering from the effects of fright, illusions, &c.; it may form part of a fixed immobile expression. Corrugation may be associated with overaction of the frontals in a similar athetoid defect, producing square creases vertical and horizontal, or in finer degree the combined action may produce a dull forehead.

When the athetoid condition is present we cannot judge of the mental state by such expression.

(47). *Orbicularis Oculi Relaxed.*—In a strong and well toned face the lower eyelid appears clean cut and well moulded, and the rotundity of the eyeball and convexity of the lower lid are seen; this sharpness is due to the good tone of the orbicularis oculi. When this muscle is relaxed and toneless the skin under the lower eyelid bulges forward and is baggy. This relaxed condition is indicative of fatigue and exhaustion, and is seen in the nerve depression accompanying severe and incessant headaches; these puffy eyes are usually symmetrical.

That the condition is muscular is demonstrated by making the patient laugh when the swollen look is removed.

(48). *Eye Movements Defective.*—Some children, when an object is held in front of them and then moved, follow it, not with the movement of the eyes, but with the head, keeping the eyes fixed. In other cases there are restless uncontrolled movements of the eyes; both conditions are included under this heading; the former is most commonly met with; the two conditions may co-exist.

(50). *Hand Balance Weak.*—When the hands are held out to command, the average balance, is with both upper extremities horizontal on a level with the shoulders, the hands being pronated, and the metacarpal bones and digits all in the same plane; such is the normal. In the type described as "Hand balance weak," the hand, when held out, is slightly drooped or flexed at the wrist, the palm or metacarpus slightly contracted or arched laterally, and the digits moderately flexed. The type may be varied: with less degrees of weakness the hand is as in the normal with the thumb drooped only; in exhaustion and great feebleness the metacarpus is more contracted or adducted, and the degree of flexion is greater.

A bad type is seen when children holding out their hands droop both thumbs and bring them together in the median plane.

(51). *Hand Balance Nervous.*—In this posture the wrist is slightly drooped or flexed, the palm of the hand slightly contracted, the thumb extended backwards, and the fingers at the knuckles are over-extended.

The various elements in this posture may vary in degree; the most essential element appears to be the extension backwards of the fingers at the knuckle joints, and this may affect the various fingers differently. The term used for this posture is empirical.

It is common in children with slight chorea, and in those who are the subjects of night-terrors and tooth-grinding, it also accompanies recurrent headaches.

It has been represented by artists in antique bronzes and drawings on vases, as well as in modern works, especially in female figures.

(53). *Lordosis.*—This arching forward of the lumbar spine is due to weakness of action among the spinal muscles. When a child holds out his hands the centre of gravity of the body is moved forward. In a strong child this is not followed by marked change of posture in the spine, but in a weak child lordosis may follow, often with temporary lateral curvature and unequal balance of the shoulders while the head and neck are thrown back.

(56). *Grinning and Over-Smiling.*—Grinning or over-smiling is usually symmetrical, but may be unequal on the two sides of the face.

With low class brain conditions it is sometimes seen as almost the only facial movement occurring upon any stimulus as a uniform movement, almost as athetoid in character as the frequent overaction of the frontal muscles.

Habitual grinning, and in particular the finer forms of over-smiling, often leave permanent naso-labial creases marked upon the skin; these may remain after the habit has been lost. If the skin be thin, a duplicate or triplicate naso-labial crease may be formed; this is more common in neurotic than in imbecile subjects.

(57). *Mouth Open.*—The open mouth in a child usually depends upon the dropping of the lower jaw. This habitual dropping of the jaw depends upon want of tone in the temporal and masseter muscles, rather than upon spasm of the depressors; it may be called to mind that this want of tone is due to lessened stimulus of the motor division of the fifth nerve, whose sensory branches are largely distributed to the meninges; weakness of this nerve leads to open mouth, irritation of it to tooth-grinding. Of course this condition of " mouth open " is only to be looked upon as a nerve-sign when the respiratory passages are unobstructed.

(59). *Response in Action Defective.*—Dealing with groups of children in a uniform method of examination as described, it becomes easy to note the response to the word of command as seen in the action following. Response in action may be accurate or uncertain, there may be delay between hearing the command and the response; some children look at the others before responding in their movements, they seem more easily controlled through the eye than through the ear.

The response should be quick and accurate, the standard to be expected is soon learnt by a little experience. The action may be long continued, the hands of the child being held out long after the others have dropped them. There may be want of impressionability to the stimulus of the command, which may have to be repeated before the action follows ; response in imitation by sight may be, and often is, much better than that following the word of command. There are some children in whom the sound of a command may be followed by a number of irregular movements, whereas an indication through the eye, by a gesture of command on the part of the inspector, is quickly followed by accurate and good response.

(60). *Speech Defective.*—Defective conditions of palate are consistent with good speech, an impediment is not usually the mechanical effect of the form of palate. It does however often happen that with defect of speech we find an arched or a narrow palate with co-existent cerebral feebleness.

The speech of children is very important; it may be almost absent, or accompanied by stammering or impediment. On putting a question it may be long before the reply comes, the question may be repeated without further reply ; speaking to the child may be followed by a large number of irregular movements and asymmetrical postures—awkward action—but not by a verbal reply.

Binary Defects in development as to their numbers and co-relation with Mental Dulness, Abnormal Nerve-signs, and Low Nutrition.—The defectiveness in the make of a child is more strongly indicated when two malformations are present than with one only. The cases of binary defects have been arranged in Table XI, which shows that the frequency of occurrence of the combinations differs greatly. The number of combined defects registered was : Boys 1240, Girls 683; total 1923. (Cases 1888-91 only).

The number of children presenting binary defects has not been determined; some children presented more than two defects so that their number must have been less than the number of cases given above, probably not two per cent of the children had more than two defects.

A fair estimate of the co-relation of the combined defects may be given in percentages.

With Mental Dulness	..	..	.. 45·7.
With Low Nutrition	..	..	.. 31·0.
With Nerve-Signs	..	..	.. 60·3.

This co-relation is higher than for single defects.

Information is then given as to the co-relations of Defects in development of the body as they occur, singly and in combination.

The statistical records of the cases seen 1888-91, and of cases seen 1892-94, are in the hands of the present Committee, and have been used as the basis for this Report. Figures relating to each enquiry are given in the various Tables.

CHAPTER V.

NOTE BY Dr. FRANCIS WARNER

On the scientific value and significance of the signs employed.
The Cases here referred to were seen in 1888-91.

Commencing my research by the study of physiognomy, cranial form and conditions of development of the body, it soon became obvious that these points were too far removed from the direct evidence of mental status such as it was desired to obtain. Indications of the individual's status in development are signs of his congenital make rather than of his culture; signs of habitual and actual present modes of working of the nerve system were needed. It then became necessary to study brain conditions by direct observation of nerve signs expressing them.

In the methods ordinarily employed in clinical study of the nerve system, movements of various kinds are the principal signs looked for and recorded; we note such actions as the patient may perform and we experiment by applying certain stimuli to the nerve centres, by examining reflex actions, by electrical tests, and by movements imitated through the eye or performed to the word of command. All expression of mental action is by movement, whether in speech, gesticulation, and facial expression, or by writing, which is a result of movement. It is by logical analysis of the neural action corresponding to such visible movement that we may hope to demonstrate the kinds of nerve action which correspond to mental states. My former endeavours in this direction have been published ["Mental Faculty," Cambridge University Press]. In seeking to arrange a set of signs available for the present enquiry it was then in motor signs that the work had to be advanced. In 1879 I was able to publish a few more nerve signs,* and in 1883—87 described apparatus for recording movements by graphic methods; the apparatus is now in the South Kensington Museum.

Tracings of movements are permanent records capable of analysis; they indicated the part moving and the time, and frequency of action in the nerve-centres corresponding, and also rendered it possible to record the conditions antecedent to such action.

While studying visible movement it became obvious that certain

---

* Brit. Medical Journal, Dec. 6th. Journal of Physiology, Vol. IV.—Vol. VII.

typical postures often correspond with definite and definable physiological and pathological conditions, and might, therefore, be used in recording such states. Such observations and lines of study lead to the enumeration of new clinical signs; we shall observe the part moved, the time and the quantity of action, and where possible the antecedents and sequents of the act. The value of such nerve-signs lies in the fact that they indicate to us the action of nerve-centres or loci of nerve tissue, thus affording evidence of action in the brain mass. The part moving and the direction of its movement are of course described in anatomical terms, the time and the quantity of a motor act are its observable attributes, the former is seen during the act, the quantity is often best estimated by the balance or posture of the parts resulting from the movement—hence we study movements and postures.

The antecedent of the movement is important to the value of the observation, hence in the school enquiry great care was taken as to uniformity of method and of word of command in viewing the children. Analysis of movements enables us to give definite nerve-signs; it is in terms implying combination and series of movements that general nerve-states are most conveniently described. Movements are often observed without any known circumstances stimulating them— such are seen in the infant's movements in uncontrolled movements of eyes and in finger twitches. The movement of a single part may be uniform in time and quantity, or it may vary. When action in several parts is noted we have a combination and series of combinations of acts, making up a complex phenomenon—such series may be classified as follows, and the nerve-action corresponding may be indicated.

1. Uniform series of acts.
2. Diminishing series of acts.
3. A series of acts adapted by circumstances.

A uniform frequently repeated or athetoid series of acts is seen in the overaction of the frontal muscles and corrugators to be described, as well as in some forms of defective expression and in tremor. It corresponds to action of a certain group of nerve-centres in repeated, often uncontrolled action.

An augmenting series of movements, corresponding to a spreading area of nerve-action may be seen in a spreading smile or facial expression, or a burst of laughter, and in the march of movement as from face to head and hand—in protrusion of the tongue on any stimulation; in the head held on one side when any question is asked, and in the fidgetty fingers of the examinee. Such spreading action is antithetical to good intellectual function.

A diminishing series of acts with lessening of the area of motor cells in activity, is seen in the child who is getting quieter after some excitement.

In action adapted by circumstances, we have a high-class function commonly called co-ordinated action, and if the co-ordinating conditions were some time antecedent, the action is considered more strictly mental in character.

Let me describe the facts seen in an infant, and make deductions therefrom. In a healthy new-born infant we find movement in all its parts, while it is awake, *i.e.*, while its brain is in full functional activity. These movements may be seen in the limbs, especially in the digits, which may move separately; they are slower than most of the movements in adults, they are almost constant and are but little under control of impression through the senses. Such spontaneous movement I described under the term Microkinesis in 1888.* When the infant is about three months' old we may observe some control of its movements through the senses. The microkinesis remains as the marked character, but the combination of nerve-centres acting, are to some extent co-ordinated by sight and sound. At the age of four or five months further evidence of control of the centres through the senses is seen; the sight of an object may temporarily inhibit the movements, and this may be followed by turning of the head, eyes and hands, towards the object seen, *i.e.*, the co-ordinated movement occurs sequent to a period of inhibition of spontaneous action following stimulation. We infer from such observations that at birth the nerve-centres act slowly and independently of one another, and the time and order of their action may be temporarily suspended by external stimuli, and during the time when no efferent currents are passing from them to produce visible movements, they undergo a change, subsequently indicated by new and special co-ordinated movements.

This appears to be a new and great advance in the infant's cerebral evolution.† When a year old, action well adapted by impressions received becomes very marked, and the child makes certain charac-

---

* Proceedings of the Royal Society, Vol. XLIV., and Journal of Mental Science, April.
† The Study of Cerebral Inhibition. Brain, Vol. LXIII.

teristic sounds on sight of certain objects; its spontaneous brain action becomes gradually more and more capable of co-ordination.

It appears that whereas at birth the most marked character of the nerve-centres is the spontaneous action of individual loci of nerve tissue in advancing evolution this spontaneity is not lost, but remains as the foundation of so-called voluntary and intellectual action becoming more controllable by circumstances. Aptitude for mental action appears to depend upon the capacity of nerve-cells for control through the senses, such impressions temporally inhibiting their spontaneity and arranging them functionally for co-ordinated action. The imbecile infant does not show this microkinesis in the normal degree, its nerve-centres are wanting in spontaneity, and later in capacity for co-ordination.

It is not my intention here to branch off into the study of physiological psychology, but it is quite possible to follow the apparent grouping of action in nerve-cells corresponding to many well-known modes of mental action. It may be shown that well co-ordinated visible movements usually accompany well controlled mental action, and a spreading area of movement not controlled often accompanies mental confusion.

This spontaneous movement, slightly under control, is the character of healthy brain action of children in the infant school, so that postures are less available as signs among these very young children, and spontaneous movements of their fingers is the normal action. The parts of the infant are then full of spontaneous movements; an exception is in the eye movements, which are not frequent in many cases. One of the endeavours of infant training should be to encourage eye movements, then to control them.

Speaking briefly of the neural action corresponding to some form of mental action, we may take the common example of imitation through the eye. Here the objects imitated are gestures or movements in another person. When the child imitates the teacher, it appears that the sight of certain movements in the teacher is followed in the child by action in the nerve-centres, which correspond to those in action in the teacher. This appears to be a result of their common inheritance.

Imitation may also be through the ear, as when the pupil repeats the question asked. Many of the children performed the movements desired better by imitation from my movements than from the word of command. Imitation may be followed by adapted action, as when after repeating the question asked, the child proceeds to answer it.

Imitation through the eye seems to be the simpler for the child than to obey a verbal command; children slow in action will look to see what the other children do and then copy them.

The postures or attitudes of the body imply balances, or ratios of action in the nerve-centres corresponding; the clenched fist or convulsive hand is common in fits and in tetany. These postures indicate relations in quantity of action among nerve-centres. If we take the 2,285 cases seen 1888-91 representing deviations from the normal balance of the hand when held out, we find that 1,029 of them presented visible defects in development also, that is to say, in nearly half of these cases with unusual or defective ratios of nerve-action the proportioning of parts of the body was visibly abnormal. This suggests the hypothesis that the forces or antecedent conditions which caused ill-proportioning of the body may also have caused a tendency to ill-balancing of nerve-centres. The converse of the proposition may be true; we have not as yet had sufficient evidence, but the suggestion may guide enquiry; it may be found that as over-action of the frontal muscles is very common with defects of the cranium, and overaction of the frontals is largely the outcome of want of mental stimuli, further culture of the mental faculties will improve the average cranial development and lessen over-action of the frontals at the same time.

In former writings* I have given a catalogue of specimens showing that in living things the time, quantity, proportions, and kind of growth may be controlled by physical forces, such as mechanical strain, heat, light, gravity, sound, &c.

It has also been shown that the distribution of abnormalities of development varies greatly in different districts; this may be owing to local circumstances. If we take the 5,487 cases with abnormal nerve-signs, we find among them 3,071, or 56·0 per cent., who also present defects in development; conversely among the 5,851 cases of defects in development, we find 3,071 cases with abnormal nerve-signs, i.e., 52·4 per cent.

If we take cases with two defects in development, such as are given in Table XVIII., we see that they are co-related with nerve-signs in percentage, varying from 43·2 up to 68·0.

The general statement that mal-proportioning in visible parts of the body and abnormal nerve-signs are often coincident, may be

---

* "Anatomy of Movement," Hunterian Lectures, 1887.

further illustrated, and such enquiry may lend some support to the hypothesis that both kinds of defects may be due to the action of physical forces controlling quantities or ratios of vital action.

In Rickets there is a marked tendency to mal-proportioning in the skeleton: this is seen in epiphyseal over growth, in unequal bilateral growth of the shafts of bone producing curvatures, and in the skull producing bosses and deformities. This tendency to mal-development may affect the features and soft parts; among 196 rachitic children 15 were small in growth, and 40 presented defects of ear, epicanthis, features, palpebral fissures, mouth, &c. This also is a condition that falls much more commonly upon the boy than the girl, about one-third of these mal-proportioned rachitic children presented abnormal nerve-signs and mental dulness.

In observing conditions of development and *physiognomy as indications* of probable conditions of *mental status,* as in older physiognomical studies, the assumption is made that visible conditions of defect in form are more or less necessarily coincident with defective brains. Such correspondence does, doubtless, often occur, but the generalization is too empirical to be applied with safety to the individual child. Here the observation of a number of abnormal nerve-signs, helps to supply the missing link and observations quoted show that among children with defects in development and abnormal nerve-signs one-third are reported by the teacher as dull at school lessons.

In the Report and in quotations from it the term "Defect in Development" is frequently used; this signifies deviation from the average or normal. I do not wish to assert that these signs are degenerations, the evidence derived from ancient works of art shows that many are of ancient date, it appears that in some classes there may be irregularities which further evolution, if wisely guided, may remove with their attendant evils.

Among the 2,961 Jew day school children, an ancient race, uniformality of development was very marked with 7·5 per cent. of deviations from the normal, and all points in nutrition, nerve-action, and mental status, appeared more regular among them than with our English children. When it is pointed out that of English day school children 10·8 per cent. and of the Jew children 7·5 per cent. present deviations from the average development, it is obvious the proposition may be put thus, the English children to a percentage of 89·2 and the Jew children to a percentage of 92·5 have evolved to an average type.

It is very usual to see disordered conditions of the nerve system in children with defective construction of body, this was the case in 3,071 children; we may see these nerve disturbances in children of normal construction of body, this was noted in the report in 2,416 children, here such signs would appear to result from the disorder produced by special circumstances rather than from defects in original construction.

In illustration, children fatigued and in the condition of chorea, may be described. Among signs of fatigue are the slight amount of force expended in movement often with asymmetry of balance in the body, the fatigued centres may be unequally exhausted, spontaneous finger twitches like those of younger children may be seen and slight movements may be excited by noises. The head is often held on one side, the arms when extended are not held horizontally, usually the left is lower, the hand balances in the weak type of posture often again more markedly on the left side.

Facial expression is lessened and the orbicular muscles of the eyelids relaxed, leading to fulness under the eyes, while the eyes themselves fix badly.

The purpose in view is to show that we may give descriptions of children in terms connoting signs observed even when we are recording indications of brain-states indicative of potentiality for mental action. A description may be given of a typical case of what is commonly called a " nervous child," such children are apt to be irritable and passionate and often suffer from headaches and hacking cough without lung disease. Let the hands be held out, with the palms downwards, and the fingers separated. The left upper extremity is often at a lower level than the right; the "nervous hand" is seen on either side, perhaps more marked on the left. There may be finger twitching, separate digits moving in flexion and extension, or laterally in adductor and abductor movements. The spine is arched too forward in the lumbar region, often with inequality in the level of the shoulders and slight lateral curvature. The face as a whole is usually too immobile, although there may be some over-action of the muscles widening the mouth on one or both sides. The tongue when protruded is too mobile.

The eyes move, mostly in the horizontal direction, their movements not being fully controlled by the sights and sounds of the objects around, except under strong stimulation. The head is sometimes partially flexed with inclination and slight rotation towards the same side.

Some of the teeth are usually found ground at their tips; this is most commonly the case with the canines. This grinding action is produced by the masticatory muscles during sleep, and is owing to irritation of the fifth pair of cranial nerves. We may here call to mind the fact that the sensory division of the fifth nerve is distributed to parts inside the skull as well as those outside it.

The individual signs suggested for observation in this enquiry have been given, but it is convenient here to make some remarks upon the parts of the body whose movements and balances in action are the most expressive of special nerve conditions.

The face is the most accurate index of the action of brain.

It is convenient to divide this region into three zones, the frontal above the line of the eyebrows, and a middle zone separated from the lower by a line at the level of the lower margin of the orbits.

The greatest degree of expression is, I think, seen in the frontal region, mainly produced by action of the frontal and corrugator muscles.

In looking at the mid-zone of the head and face, the observer's eye traverses it from ear to ear, noting these features, the palpebral fissures and the tone of the large orbicularis oculi muscles, the bridge of the nose both in its bone and soft tissue, as well as the eyeballs and their movements.

Signs are given for each of the three zones.

The hand when held out free and not mechanically restrained, affords the next most important index by its movements and the balance of its parts; I have described eight typical hand postures, but found in practice that two were sufficient for the present purposes.

CHAPTER VI.

THE GROUPS OF CHILDREN AND GROUPS OF CASES.

We proceed to review the children as groups of cases arranged according to their mental and physical status; as we proceed many children must be shifted from one group to another. As far as possible the groups specially referred to are given in the same order as in the catalogue (see page 82), in which the groups are enumerated, and to which reference numbers are appended. For the purposes of comparison and remark, it has been found convenient to bring together the dull children and the afflicted children.

GROUP 1—*Normal or Average Children.*

Such children are the average as presenting no visible defects or abnormal conditions in development, in nerve-signs, or in ordinary work with the teachers in school. These children were passed over, and no notice was taken of them, beyond their number and distribution in the school standards.

It would be interesting to review these children for the purpose of determining their relative points of excellence; this could be done, taking as the standard points of excellence, and using anthropometric and nerve-tests indicating superior development and perfection of nerve-action. Scholarships and the means of higher education might with advantage be given to the children of best physique and brain power. It is seen here, as elsewhere, that the girls take a general precedence of the boys in freedom from defects during school life.

GROUP 3—*Eye Cases.*

No tests were used as to acuteness of vision, or errors of refraction, but when the eyes were looked at obvious defects were noted. Cases of ophthalmia were not registered, but some of its late effects were recorded under the headings "Disease of Cornea" and "Eyes lost by Disease." Ophthalmia was seen in several day schools. The number of children said to have lost an eye by accident seems to be large; in one school two brothers had each lost an eye from playing with toy guns.

Under the heading "Squint," cases of organic strabismus requiring operation, as well as examples of varying strabismus, are recorded; probably many are instances of hyper-metropic strabismus that might be corrected by spectacles.

Children using spectacles in school were enumerated among " Eye Cases," and as errors of refraction are known to be very common, a majority of them must have been undetected in this inspection, it follows that a small number of eye cases here noted in a school does not necessarily indicate the presence of but few cases requiring attention. The group of children registered as " Eye Cases" demonstrates what a large amount of ophthalmic work is needed among children, and the fact that with 1,622 cases of Squint, only 644 children had convex glasses, shows that spectacles must be required by many who do not use them.

GROUP 4—Cases of Rickets.

1888—91 .... Boys 157. Girls 39. Total 196

Probably more children were or had been rachitic than those registered ; when the conditions seen in the bones left no doubt the case was registered accordingly, but the body could not be examined in detail under the conditions of this inquiry. It seems that a great characteristic of the conditions termed rickets is the malproportion of growth in the skeleton, especially about the cranium. It is shown that the palate is frequently ill formed, and also that defects in development other than cranium and palate were found. The pathological question might be raised whether a large proportion of the cases registered as " Cranial Bosses," a sub-group of the cranial abnormalities, were not really cases of rickets. These rachitic children are badly proportioned ; among them were as follows :—

Conditions presented by the Rachitic children.

Boys.	Girls.	Total.	
143	25	168	Cranial abnormalities arranged below in sub-classes.
79	14	93	Head large and ill-formed.
51	9	60	Cranial bosses, principally frontal and usually symmetrical.
8	1	9	Forehead mis-shapen.
3	0	3	Head asymmetrical; sometimes one frontal boss only developed.
1	0	1	Dolicho-cephalic.
1	1	2	Head small.
23	8	31	Palate defective in form.
21	6	27	Cranium and Palate defective.

Boys.	Girls.	Total.	*Condition presented by Rachitic children* (cont.)
58	11	69	Defects in development other than cranium and palate.
32	13	45	With indications of low nutrition.
54	15	69	With abnormal nerve-signs.
64	10	74	Reported as dull by the teachers.

Of these Rachitic children 69 presented defects other than those analysed above, they were as follows:—

	Boys.	Girls.		Boys.
Small in growth ..	10	5	Features coarse	3
Defect of Ear	23	2	Palpebral fissures small ..	2
Epicanthis	6	2	Mouth small	1
Deaf..	2	1	Forehead hairy	1
Nasal bones wide ..	4	0	Congenital defect of hand..	1
Prognathous	1	0	Congenital defect of eyes ..	2
Frontal ridge	1	1	Epilepsy	1

GROUP 28—*Cases presenting Defects in Development.*

Conditions of mal-development stand as the largest class of visible defects observed; their co-relation is high, and as signs easily recognised and capable of description and classification, they stand prominently forward as pathological conditions characterising portions of the child-population as deviating from the average or normal.

The importance and the difficulty of defining and maintaining a fixed standard of defectiveness or deviation from the normal has been referred to. Analysis and comparison of cases show the signs of defective development observed to be of different value and significance, to demonstrate this the co-relations of each sign have been determined; from the point of view of estimating the potential mental capacity, these signs are only of value in as far as experience gained by observation shows their average co-relations with cerebral or mental defects.

Defects in development mainly indicate congenital and constitutional conditions or results of inherited impressions, perhaps in part due to the environment of the parents, with a pre-disposition to delicacy both in body and in brain-action; they are very frequently met with in all classes of society, not least so among the upper social grade. (See Tables XVI., XIX., XXVI., XXVII., and XXVIII.)

It appears probable that to a great extent, such defects may be rendered less numerous among the population by hygienic care with regard to buildings, light, and air.

If State medicine is to make any effort to remove these faults in development among the population, descriptions of groups of children must be given in terms indicating their bodily condition, and defects in form and proportions will probably first attract the attention of scientific workers in this direction.

If we could discover the causes of mal-development and remove them, we should have fewer children with abnormal nerve-conditions, low nutrition, and mental dulness. It appears that one means of studying the causation of physical defects may be by observing their local distribution. The cases seen in 1888—91 distributed in 20 localities and in certain groups of schools were given in a former report.*

It should be distinctly stated that a defect in development may not be accompanied by any defective condition in the individual, yet the case may serve us in giving some clue as to the causation of such defect; for this reason each defect, or rather its absence, was looked for in each child seen. Children presenting one, or even two defects, are by no means necessarily in any further sense abnormal children, or exceptional from an educational point of view. Cases of mal-development form a good standard for determining the material in a school.

GROUP 29—*Cases Presenting Abnormal Nerve Signs.*

This group of children is nearly as large as that presenting defects in development, this fact and its co-relations demonstrate that the observation of movement and balances of the parts of the body may give important indications of the conditions of children.

This group of signs indicates the activity and balance of action of the nerve centres; the conditions thus indicated may be temporary; the observer can but note what he sees and repeat his observations. In some cases the postures assumed are in imitation of the teacher, or have been taught in the school, and many defective actions are clearly removable by careful training and well regulated exercises.

These defective signs indicating as they do brain status much associated with mental dulness, may in most cases be removed by good training—here is an outcome of this work presenting a hopeful aspect and one worthy of earnest attention on the part of educators. It is of more importance to have a child's brain in good working order, and well trained and brought under control, than that he should acquire any special knowledge without this. The purely physical aspect of the case is that all abnormal nerve signs are largely co-related with

---

* Journal of the Royal Statistical Society. Vol. LVI. March, 1893.

defects in development, that is to say, some mal-proportion in the parts of the body is largely associated with a tendency to ill-balance among the nerve centres.

The significance of these signs varies in two directions; some indicate an over mobile nerve system, the centres tending to separate and spontaneous action, not well under control through the senses, of which finger twitching is the type ; and a second set which indicate low class brain development, these are mostly repetitive uniform movements, athetoid in type and represented by chronic overaction of the frontal muscles and repeated grinning. Much further analysis as to the grouping of these signs is needed to elucidate their full significance. Others again indicate general low nerve power specially relaxation of the orbicularis oculi, the weak hand posture and lordosis. The type commonly called a nervous child is characterised by lordosis, the nervous hand posture and finger twitches; such children are often mentally bright and these signs have the lowest pathological co-relation.

Of all signs the tone of the orbicularis oculi is the earliest sign of commencing fatigue, a relaxed and toneless condition may be due to many causes, such as late hours the previous day, mental exhaustion or ill-ventilation. It is very common as an accompaniment of head-aches.

As to the children presenting irregularities in action of the nerve system, " Nerve cases," their careful training in school may do much to prevent them from growing up permanently nervous or mentally dull ; under unfavourable training the proportion with nerve signs and the proportion with mental dulness rises.

GROUP 30—*Cases Presenting Low Nutrition.*

· Low nutrition was recorded against any child seen to be pale and thin or delicate; it is not sufficient to look at a child's face, the limbs should be felt to ascertain that the child is not thin in the body, though fat in the face as is commonly the case with nervous children, in whom the face is often the best nourished part. No enquiries were made as to the feeding of these children, but it may be assumed that in the resident schools, and among the children attending " the twenty better class schools," food was supplied in sufficient quantities.

The most obvious fact concerning the children of low nutrition is that a large proportion of them presented some defects in development, whether in day or resident schools, or in those of upper social class.

Here then we see one of the ill effects arising from defects in development.

Life in a resident school greatly reduces the amount of low nutrition among the cases of mal-development. (See Table II., page 14.)

In the boarding school there appear to be two conditions at work, (1) the regular feeding, sleeping, &c.; (2) uniformity in living, and freedom from the many impressions and troubles of the less protected life at home and day schools, where the child has its life at home, life in the streets and in school, often with late and irregular hours of sleep.

TABLE III. *(seen 1888—91)—Showing Number of Children with Low Nutrition in Resident and Day Schools, and the Percentage of these Cases that presented co-related conditions taken upon the Cases of Low Nutrition.*

Number of Cases of Low Nutrition.	Percentage of Development Cases.			Percentage of Nerve Cases.		
	Boys.	Girls.	Total.	Boys.	Girls.	Total.
Resident Schools, I.-XXXIV.— B. 291. G. 156. T. 447............	68·3	85·2	74·2	66·6	69·0	67·3
Day Schools, XXXV.-CVI.— B. 739. G. 817. T. 1556	72·2	72·5	72·4	59·6	60·0	59·8
For all Schools, I.-CVI.— B. 1030. G. 973. T. 2003.........	71·1	74·6	72·8	61·6	61·4	61·5

Number of Cases of Low Nutrition.	Percentage of Nerve Cases with Development Defects.			Percentage of Cases reported as Dull by the Teachers.		
	Boys.	Girls.	Total.	Boys.	Girls.	Total.
Resident Schools, I.-XXXIV.— B. 291. G. 156. T. 447............	40·8	48·7	43·6	33·3	54·4	40·7
Day Schools, XXXV.-CVI.— B. 739. G. 815. T. 1556	39·6	37·3	38·4	41·2	37·9	39·5
For all Schools, I.-CVI.— B. 1030. G. 973. T. 2003.........	40·0	39·1	39·5	38·1	40·5	39·7

GROUP 38—*Cases with defects in development, Abnormal Nerve-Signs and Low Nutrition.*

This group appears to represent a special class of development cases, in which the inheritance has produced not only

visible mal-formations or ill-proportioning in the body, but also a
constitutional tendency to low nutrition and a state of nerve-centres,
ill-balanced and acting badly. Such children may be said to be
delicate, and in 1888-91, 44 per cent., and in 1892-94, 50 per cent. of
them were reported as dull.

The children when dull are included in Group 12, who appear to
require special care and training.

GROUP 42—*Children presenting development defects without Abnormal
Nerve-signs.*

These development cases with a well-regulated nerve system
present less mental dulness than those with nerve-signs, showing
the importance of studying nerve-signs.

Of development cases with nerve-signs (1888-91) 43 per cent.,
(1892-94) 49 per cent. were dull.

Of development cases without nerve-signs (1888-91) 35 per cent.,
and (1892-94) 37 per cent. were dull.

GROUP 45—*Children presenting Abnormal Nerve-signs without defect
in development.*

Here no defect in development accounts for the nerve-signs,
they appear due to other causes; they are slightly more frequent
among the resident children, and among the upper classes than in
the average day schools, so that low feeding does not appear as a
necessary cause. It is probable that in this group we have children
of normal make who are ill-trained, neglected, and over-pressed by
the stress of life. These seem to be the children most improvable by
altered conditions and appropriate training; many of them are dull.

GROUP 55—*Children presenting no defects in development or Abnormal
Nerve-signs, but reported as dull by the Teachers.*

Such cases present good physical development, and a sound
condition of brain as indicated by motor action. It appears that the
brains of these children, though capable and healthy, had but little
power for school work. It is important to differentiate such pupils
from those with defective conditions.

The Group is small and includes some children noted as crippled by
disease, epileptic, with eye defects or low nutrition, and the cases
presented by the teachers as dull in whom no defects were found.

The mental examination and history of some of these cases is important, as it may show grave defects in moral sense and intellectual
power, unobservable by simple physical observation.

GROUP 31—*All Dull Children.*

The children reported by the teachers as mentally dull or below the average in ability for school work were mostly selected by the signs observed before the teachers presented any who were passed over by the observer, and the teachers generally acknowledged that the dull children had been so selected. This fact in itself shows the strong link existing between physical conditions, including nerve signs, and the causes of mental dulness. Tables XVI. and XVII. show the number of cases presenting certain conditions, the proportion of these mentally dull, and the percentage of the latter, taken upon the number presenting the conditions respectively.

In the facts here given it is seen that defects in development and abnormal nerve signs are largely co-related with mental dulness. The brain disorder so largely accompanying defect in development is also largely productive of mental dulness, it is to this fact that attention is particularly directed. The nerve-signs are an index of the brain condition, and methods of training which remove the nerve-signs tend to produce a brain condition with better aptitude for mental work. These nerve-signs can be dealt with in detail, and school training may be adapted to removing in turn each bad balance of body, while physical exercises are used for the cultivation of quickness and accuracy and full power of eye movements, etc.

Children "feeble-minded," or exceptional in mental status.

This is one of the most important classes of children, though happily it is small in number. As the research proceeded further analysis was made concerning these cases, which will therefore be presented separately for the two enquiries made. (See Tables XIV., XXIX., XXX.)

TABLE IV. (*cases seen* 1888-91)—*Children " feeble-minded," or exceptional in mental status, their coincident defects.*

	Boys.	Girls.	Total.
Number found in schools	124	110	234
Of these children the following presented coincident defects:			
Cases defective in development	84	75	159
Abnormal nerve-signs	96	77	173
Nutrition low	26	25	51
Development defective with abnormal nerve-signs and low nutrition	17	13	30
Epileptic	5	5	10
Crippled or paralysed	5	3	8
Eye cases	20	16	36
Hands blue and cold...	2	3	5

In one National school and also in one Poor Law school there were three cases from one family.

It is difficult to define what physical conditions seen, as apart from mental signs, indicate the child unfitted for the average methods of training, and any arbitary attempt to do so must fail. For distribution of these cases in Divisions of Schools, see Table X. This afflicted class includes idiots, imbeciles, children feebly-gifted mentally, and children mentally exceptional; as to definitions and numbers (see Catalogue, page 82), 174 are entered as feebly-gifted mentally, i.e., of defective mental capacity short of actual imbecility; it is probable that some of these children would be found on further examination to be imbecile, and many may be capable of great improvement.

Cases seen 1892-94.

Children "feeble-minded," or exceptional in mental status, include Groups 6, 7, 8, 9. They are presented in Tables XXIX. and XXX. arranged as to standards and ages, and their distribution into primary groups of defects; a few comments may be added as to each group.

GROUP 7—Imbeciles.

Boys 3. Girls 2.

None were epileptic; 1 girl was paralysed. 2 boys had cranial defect, 1 defect of palate, and 1 nystagmus. Neither of the girls had defect of cranium, palate, or eyes.

GROUP 8—Children feebly gifted mentally.

Boys 49. Girls 52. Epileptic: Boys 3. Girls 5. Crippled: Boy 1. Girls 2.

16 boys had defect of the cranium, and five defect of palate.

8 boys had eye defects, including 3 with nystagmus.

23 girls had defect of cranium, and 7 defect of palate.

5 girls had eye defects, including 1 with nystagmus.

GROUP 9—Children mentally exceptional.

Boys 4. Girls 3.

None were epileptic or crippled; there were no eye cases.

1 boy had defect of cranium; there was no boy with palate defect.

1 girl had palate narrow; none with defect of cranium or eyes.

The experience of hospital physicians and philanthropic societies shows that neglect of feeble-minded children of all grades leads to much social evil. The blind and the deaf are happily now cared for under the provisions of the Elementary Education Act (Blind and Deaf,) 1893, and teachers are specially trained for this work; but except in a few centres the children of the various grades of feebleness, short of imbecility, children who present a deficiency, are in many schools unwelcome, and no encouragement is given to the school authorities to collect or care for them, they are an incumbrance if not properly provided for, and untrained, they tend to social failure, pauperism, and criminality.

GROUP 10—*Epileptics, and Children with History of Fits during School Life.*

Seen 1888-91.	Boys 36	..	Girls 18	..	Total 54.
With defects in development..	„ 19	..	„ 9	..	„ 28.
With nerve-signs	„ 22	..	„ 13	..	„ 35.
With low nutrition..........	„ 6	..	„ 5	..	„ 11.
With mental dulness	„ 23	..	„ 12	..	„ 35.

These cases were enquired for in every school, and in some instances children not attending school were sent for by the teachers. Any case with a history or indications of fits during school-life was recorded for what it may be worth. A list of these cases has been published.* It would appear that most epileptic children are absent from school. Of the cases given, 5 boys and 5 girls were mentally defective.

In 1892-94, 21 boys and 35 girls were registered in this group : Crippled—no boys, 1 girl. 3 boys had cranial defect; none had defect of palate. There was 1 eye case—a boy. 2 girls had cranial defect, and 1 defect of palate. 1 girl, eye case.

GROUP 11—*Children Crippled, Paralysed, Maimed, or Deformed.*
(Not Eye Cases.)

Seen 1888-91.	Boys 155	..	Girls 84	..	Total 239.
With defects in development ..	„ 44	..	„ 27	..	„ 71.
With nerve-signs	„ 51	..	„ 22	..	„ 73.
With low nutrition	„ 44	..	„ 18	..	„ 62.
With mental dulness	„ 57	..	„ 36	..	„ 93.

These children varied greatly in brain power—some were mentally bright, others dull; they also varied in conditions of health. The conditions of disease causing crippling were in various stages, and

_____

* See "The Feeble-minded Child and Adult." C. O. S. Series. 15, Buckingham Street, Strand, W.C.

many of these children were capable of work and play. 5 boys and 5 girls were mentally defective.

Cripples from congenital defects..Boys 7 .. Girls 9 .. Total 16.

    „   „   disease or injury .. „ 88 .. „ 53 .. „ 141.

    „   „   paralysis .... „ 60 .. „ 22 .. „ 82.

    Seen 1892-92 .... Boys 75 .. Girls 60.

7 boys had cranial defect; 3 had defect of palate; 4 had eye defect. 5 girls had cranial defect; none defect of palate; 2 had eye defect.

The classes of cripples are enumerated in the nomenclature. (See page 72.)

DEFECT (55)—*Children Deaf or partially Deaf.*

    Seen 1888-91.   Boys 34 .. Girls 33 .. Total 67.

With defect in development .... „ 27 .. „ 22 .. „ 49.

With nerve-signs .... „ 27 .. „ 23 .. „ 50.

With low nutrition ... . „ 3 .. „ 13 .. „ 16.

With mental dulness .... „ 18 .. „ 21 .. „ 39.

Tests for hearing were not commonly used, but when a child was found to be deaf, this was recorded; children almost totally deaf, as well as others deaf and dumb, were met with among the other children in public elementary schools. The 51 children (boys 31, girls 20) in the school for the deaf and dumb are not included in the group above; the report of 1888 gives the following account of them.

DEFECT (81)—*Children Deaf and Dumb.*
Seen 1888-91.

With cranial abnormalities ...... Boys 14 .. Girls 4 .. Total 18.

With nerve-signs .... „ 15 .. „ 6 .. „ 21.

With low nutrition .... „ 5 .. „ 4 .. „ 9.

With mental dulness .... „ 4 .. „ 2 .. „ 6.

Eye cases..................... „ 2 .. „ 3 .. „ 5.

Analysis of 51 Deaf and Dumb Children.—In this school each child was examined separately, and the condition of the palate was noted.

Particulars as to head measurements and conditions of general physical health were most kindly filled in by Dr. Martyn, medical officer to the institution. The Secretary of the institution very kindly filled in some particulars of the history of each pupil from the records kept.

The ages of these children varied from 7½ to 11 years, average age 8 years. The following table indicates the cause assigned for the deafness.

TABLE V. (*Cases seen* 1888-91).—*Deaf and Dumb Children.*

Congenitally deaf	Boys	21	..	Girls	11	.. Total	32.
Sequent to measles	,,	2	..	,,	1	.. ,,	3.
,, whooping cough......	,,	1	..	,,	1	.. ,,	2.
,, scarlet fever	,,	—	..	,,	1	.. ,,	1.
,, sunstroke	,,	1	..	,,	—	.. ,,	1.
Fits or brain disease............	,,	3	..	,,	2	.. ,,	5.
Effects of a fall................	,,	2	..	,,	—	.. ,,	2.
First dentition	,,	—	..	,,	3	.. ,,	3.
Cause not stated	,,	1	..	,,	1	.. ,,	2.
		31			20		51.

In 3 cases the children were the offspring of first cousins.

In 5 cases, deafness was acquired between the ages of 3 to 6 years, in 2 cases without cause assigned. The palate was examined in all these cases. In 2 cases it was "vaulted," in 2 "narrow," in 3 " arched."

Of these 7 cases 5 occurred among the congenitally deaf pupils, and 3 coincided with other defects of the cranium. In 7 boys and 7 girls the teeth were found ground. The various conditions found in these children are given in preceding tables.

It is noteworthy that most of these children fixed their eyes on any one speaking to them remarkably well.

GROUP 12—*Children that appear to require special care and training.*

Some difficulty has been experienced in giving anything like a definition of this class of children. Eye cases and the deaf and dumb are not included, not on account of their small number but because these cases could not be specially investigated in this enquiry, and attention has been fully drawn to their condition in other reports. The group includes, "children feeble-minded or mentally exceptional," epileptic, cripples, and the "development cases with low nutrition and nerve-signs, who were reported as dull mentally." The group as thus arranged, allowing for overlapping cases, contains (see Table XIV.) 817 children (Boys 473, Girls 344), or 1·6 per cent of the 50,000. Seen 1888-91. Of the number given 165 are included on physical grounds, not being mentally dull.

The numbers of children requiring special care and training as seen 1892-94, are given in Tables XXIX. and XXX. arranged as to Standards and ages.

Of the 226 boys 218 girls requiring special care, as seen in enquiry 1892-94, 61 boys and girls, 52 are included on physical grounds, not being mentally dull.

CHAPTER VII.

THE BEARING OF THIS ENQUIRY ON THE EDUCATION AND CARE OF CHILDREN.

The State has undertaken a great work and heavy responsibilities in making the education of children compulsory upon all, and this labour has not been lessened by the recent provision of free education. Besides this educational provision for children living at home with their parents, the State has taken complete charge of large bodies of children under the Poor Law, and children with criminal tendencies in the certified industrial schools under the Home Office. In addition to these groups the care of imbeciles, the blind, and the dumb children in part falls upon the State.

One of the pleasing results that has followed the improved intelligence of the population is a material diminution of crime, yet we continue to make but little provision for the care of feeble-brained children, and exempt them on the ground of their feebleness from the education shown to be necessary for the success of the average children. Might not due care of these boys and girls help still further to lessen crime as well as pauperism, and other forms of social failure?

No work is more generally popular than that undertaken to benefit the whole people, and among no section of the population are such efforts more needed and more useful and hopeful in their results than among the children. Very much is said and written about educational methods and the care of children, but have we as yet the necessary information concerning the child-population, the classes that have to be provided for, the methods of classifying them, and the means of meeting their special needs?

Where children have already been grouped as blind, imbeciles, dumb, &c., information has been collected as to their physical condition, and arrangements for them are founded upon such experience.

As to the mass of the child-population, the 6,000,000 in public schools, upon whom we spend in taxes, apart from School Board Rates and subscriptions, £10,000,000 a year, we have but little knowledge, except such as may be gleaned from the reports of school examinations by H.M. Inspectors and the returns of births and deaths. The Royal Commission on blind, dumb, and children requiring exceptional methods of education, was satisfied upon a small amount of evidence supplied to them that among the pupils in elementary schools there were many "feebly gifted," and others physically incapable of benefiting by the ordinary education; and gave it as their opinion that many of the feeble children, partly through incapacity, in a larger degree through irregularity of attendance due to feebleness, remain practically untrained, and that they are unprovided for in our Public Educational system which, under the code, is designed for average or normal children.

It is not on such grounds only that we would urge the necessity of accurate knowledge of the bodily and brain conditions of the child-population. The great object sought by State medicine in its various divisions is to acquire such knowledge, based upon scientific enquiry, as may serve to render the population healthy, long-lived, and prosperous, while universal education is provided that all may receive a mental development, fitting each as a citizen to provide for himself, and take his place among his fellows.

Public hygiene is concerned not only to lessen the death-rate and remove disease as far as possible, but also to render the population of all classes as healthy and well-developed as possible—well-made in body and sound and strong in brain power. For such work we need something like a census of the children which it has been shown may be commenced by examination in schools where attendance is now compulsory.

If the work were carried out on a sufficiently large scale in several districts and centres of population, we might soon be able to give answers to important questions, such as, " What proportion of the population of children require special educational provision?" " What are the observable effects of higher education, technical education, town life, crowded living, large block residences, deficient light, adjacent railways, cottage residences, elevated land, drainage and water supply, adjacent open spaces and trees?" " Is poverty a fruitful cause of defects in children; are these defects local in distribution, and to what extent are they determined by local conditions?" Comparison of children in various localities would show much of the hygienic effects of town and country, the peculiarities of children respectively in northern and southern counties, and the districts where degenerative developments are most prevalent; thus their causation might be determined.

Some evidence was advanced in the last chapter that such problems are not purely speculative, and that their solution is likely to be attended with practical results beneficial to the public interests.

The children who fall to the care of the State appear to be those of more mal-development than the average of the public elementary schools ; this suggests the economy of lessening such evils.

In the application of this work to State medicine the co-observation of conditions of development and nerve signs will have their place, but probably efforts might at first be directed to lessening the mal-development.

The nerve signs are indices of the finer impressions made upon the brain, forces so slight as often to be called " moral influences," the sound of words, their tone, sights imitated, &c. Badly made heads and palates are largely coincident with frontals over-acting, causes increasing the latter may be connected with the causation of the development defects. To the teachers and others in charge of children we must principally look to remove defective-nerve signs, while we try to discover the cause of defects in the child's material structure.

Among points for investigation may be suggested the study of the environment of different classes of children, under what conditions is each defect most commonly met with; the effects of environment apart from heredity might be sought. Thus we might study the effects of diminished light, railway noise, deficient exercise.

The State becomes heavily burdened by the defectively made portion of the population, which probably tends to accumulate under extensive emigration, leaving with us the weak, tending to pauperism, starvation, vagrancy, and crime ; a large body of the " unemployed " and others capable of earning only small and varying wages; the field for recruiting the services is also limited. Were this lower stratum raised it would pave the way for social improvement, higher education, better wage earning power, and less social failure. It must be remembered that these feebly-gifted children are confined to no social class, and appear to be not less numerous in the upper grades.

These observations show the harm that probably arises from exempting the feebly-gifted and defective children from all educational training because they are unfitted to compete in the school with those of average capacity. It is as important to know the average condition of the children, as the rate of mortality in any given locality. Mal-development has been shown as far as the facts go to be a potent factor predisposing to both mental dulness and low nutrition. It is not solely for the purpose of attaining a condition of the people with a smaller percentage of badly made heads, palates, ears, noses, or other bodily defects, that a strong effort is called for. In removing the causes of such defects we may hope to lessen the average of co-attendant " mental feebleness " and " low nutrition."

The ends which it is desired to attain are to improve the average development, nutrition, and potentiality for mental faculty, and thus to lessen crime, pauperism, and social failure, by removing causes leading to degeneration among the population, and by encouraging the means of improvement.

It is desirable to obtain a normal of the child-population such as is required in all branches of vital statistics.

In seeking information concerning the mortality of a town we compare its rate of mortality with the average, and if it be high we seek for the cause, so we may enquire into the average development of children in a town or district, and where the conditions are bad try to find a remedy ; we need to determine an average or normal upon sufficiently extended observation to enable us to say what groups of children or districts are above or below the average.

A large number of children must be examined under different circumstances of life and in different districts to afford any satisfactory evidence upon such points, as for example, " children with small heads ; " at least 30,000 or 40,000 should be seen in each district.

Crime in childhood is an unnatural action, and suggests in all such cases an abnormal condition of the child, so that a physical examination should be made. The National Vigilance Association have shown that a large number of those with whom they deal in workhouses and elsewhere are feeble-minded. This strongly suggests that the best means for preventing " ill-development " and " feeblemindedness " are also means of preventing crime and social failure. It seems not unlikely that many of those who wander homeless are defective. Crime may probably be lessened by early care, while exemption from even ordinary discipline and culture tends greatly to aggravate the congenital tendency to failure.

Co-relation or the Relations of Physical and Nerve-signs to Low Nutrition and Mental Dulness.

In Tables XV., XVI., XVII., the number of cases presenting each sign has been shown, and the co-relation has been added showing the number of cases presenting the sign who were also registered as presenting "low nutrition, abnormal nerve-signs, mal-development, or mental dulness" respectively. These co-relations are also given in the form of percentages in the Tables referred to.

It is not wished to represent the percentages as having an absolute value of co-relations to the sign which may be applied to an individual child. The co-relation for some signs is probably of small value on account of the small number of cases observed, but it is given as illustrating that each physical or nerve-sign has a co-relation with mental dulness, nutrition, &c. When the co-relation is on a small number of cases, the need of further observations in this direction is indicated.

The percentage form is useful as indicating that some generalisations, drawn from large groups of cases, apply equally to the individual signs characterising the group. Thus defects in development have a higher co-relation with abnormal nerve-signs in boys than in girls, but as to nutrition and mental dulness the girls suffer most.

As a contribution towards the etiology of defective development, a Table was prepared (see paper in the *Statistical Journal*, March, 1893, by Dr. Warner), distributing these cases in twenty Metropolitan districts. It is shown that the distribution is very unequal, being high in the Western district of Kensington and Chelsea, viz., 12·5 per cent. on the numbers seen, and lower in the poorer schools of Islington, viz., 7·5 per cent. This table also gives the percentage distribution of the principal defects registered, taken in two ways : (1) Upon the number of children seen ; (2) Upon the number of development cases. If such observations are confirmed by further experience, this method of arranging the facts may afford evidence upon the causes in the district tending to produce defects in development, and possibly for determining the particular kind of defect most prevalent.

In certain districts the ratio of boys and girls presenting the same conditions is not the average. As a preliminary to determining the means that may be used to try and lessen the physical causes of mental dulness, Tables are given of the conditions and groups of conditions accompanying it, showing, as far as the present work goes, the co-relative value of each sign in development and nerve action observed, as well as the distribution of these signs in certain areas or districts, in certain classes of schools and in the nationalities.

In cases seen 1892-94 the percentage of mental dulness rises from 38 for development cases without nerve-signs to 48 when both are present, reaching 52 when the mal-development is accompanied by low nutrition and abnormal nerve-signs.

Development Cases considered in Relation to Sex and Residence.

If we take 100 boys and 100 girls with defects in development, we shall find many of them with abnormal nerve-signs, low nutrition, and mental dulness. Following the experience gained, the following estimate may be given showing the probable results of placing them first in a day school and then in a resident school :

TABLE VI. (*Cases seen* 1888-91).—IN THE DAY SCHOOL.

Boys' Side.—100 boys with defects in development.	*Girls' Side.—100 girls with defects in development.*
Nerve cases 50	Nerve cases 47
Nutrition low 23	Nutrition low 38
Reported dull 38	Reported dull 40
Cases of nutrition low, nerve-signs, or dull .. 111	Cases of nutrition low, nerve-signs, or dull .. 125

TABLE VII. (*Cases seen* 1888-91).—IN THE RESIDENT SCHOOL.

Boys' side.—100 boys with defects in development.	*Girls' side.—100 girls with defects in development.*
Nerve cases 62	Nerve cases 52
Nutrition low 16	Nutrition low 20
Reported dull 40	Reported dull 44
Cases of nutrition low, nerve-signs, or dull .. 118	Cases of nutrition low, nerve-signs, or dull .. 116

It is thus obvious that residence contrasted with home life and day school produces marked effects, different among boys and girls. On both sides of the resident school nutrition becomes higher, more markedly with the girls.

Nerve-signs increase with residence, especially with boys.

Mental dulness increases with residence slightly, more so among the girls.

The loss and gain from putting 100 boys and 100 girls with defects in development in resident schools may be represented thus :—

	Boys.	Girls.
Fewer cases of low nutrition ..	– 7	–18
More cases of abnormal nerve-signs	+12	+ 5
More cases of mental dulness ..	+ 2	+ 4

The methods of procedure and report that have been explained afford useful results and might be enlarged as follows :—

1. A body of vital statistics indicating the average condition— the normal at the present time— of the school population; variations therefrom in different localities, under different circum-

stances, and at different times, might become known and lead to action for the removal of defects by hygiene and training.

2. The children in a district may be better known, the cripples, the epileptic, the mentally defective, as well as those presenting low nutrition and eye defect, &c., something like a census of conditions (as apart from surroundings) is obtained.

3. A large body of collated observations would enable a complete estimate to be made of the value of each sign and condition, this has already been done in part.

4. Information would be obtained as to causes affecting development of body and brain power, nutrition and the relations of abnormal nerve signs to mental dulness, while the varying child material in schools and results of training might be studied. With such a collection of information as that here arranged, problems can be defined and illustrated and methods for their solution may be found.

Should a public school be managed with the purpose of producing the best educational results, i.e., the greatest number of well trained intelligent children ; or as a means of conferring the greatest benefit on the neighbourhood in preventing the evils of non-education and minimizing the number of subsequent failures in life ? Probably a few higher grade schools have the former function as their duty, while the largest number of local schools have the latter and more difficult task to perform.

A School Board has both functions to fulfil. The Higher grade school may then select the best made pupils, those with the best faculty for benefiting by intellectual training. The Local school has to do the best it can for all and each child. The practical difficulty arises with regard to the feebly-gifted, the weak, the defective children, those whose physical health and brain power are below the average for which arrangements are made.

It has been shown that these children may be known and a list of them prepared for the managers. It is thought by many that while our public system gives direct encouragement to teachers to show the best intellectual results with the average and better class children, no public approval is bestowed upon the special attention they give to the feeble, deficient, and lowly-gifted.

These children should be kept under special notice and a premium or special commendation might be given for the regularity of their attendance, and for care and arrangements adapted for their improvement.

Constitutional Differences between Boys and Girls, and their Relation to Educational Requirements.

How do boys and girls respectively bear the effects of their environment ?

The first fact to be observed is that more boys than girls appear in most of our groups of cases presenting some defective conditions — an exception is in relation to low nutrition. (See Catalogue, p. 82.)

This exceptional fact suggests further analysis. In a school of 1000 boys and 1000 girls, according to the average (cases seen 1888-91) there will be children with low nutrition as follows :

	Boys.	Girls.
Low nutrition with development defect	27	31
„ „ without „ „	11	11

It is seen that without defect in development girls do not appear to be more delicate than the boys; but those with development defects are much more delicate. Due care of the "development" cases might prevent the manifestation of insomnia, hysteria, and other troubles in the girls' school, while the normal children may work hard.

Let us compare the boys and girls from the point of view of the teacher's experience—they soon find out who are dull at lessons.

Among 1000 boys and 1000 girls the average number is :

	Boys.	Girls.
Dull	82	63
Dull and delicate	14	17
Dull, with nerve-signs	51	38

Among the dull boys there are fewer who are delicate, more with nerve disorder; and physical training is more likely to lessen the proportion of dull boys than girls.

The Child Material in a School.

In any State school for boarded pupils a medical certificate is required, and applicants suffering from disease, the epileptic and those with brain defects, are refused admission and have to be provided for elsewhere.

Children can be exempted from attendance at public elementary schools on similar grounds by means of a certificate from any medical practitioner, but these rejected cases are not provided for educationally, probably their names are removed from the school register if permanent cases.

Teachers are well aware that their educational results depend in great part upon the child material admitted into the school; when the school places are not full they are not free to select applicants for admission; but this does not apply in all cases.

The proportion of ill-made brains appears to vary much in different schools, and probably varies in different localities; in a school designed to represent the requirements of a neighbourhood, it seems that allowance should be made in assessing results for the average material placed under the care of the teaching staff. A high percentage of defects in a district calls for special attention on the part of the Sanitary Authority, and also for special care in training the children, while a lower standard of intelligence must be expected. Favourable opinion was formed concerning some of the endowed parochial schools where special attention was given to the requirements of the neighbourhood, one or two epileptics, one or two feeble-brained children who could only be trained and not presented for examination were allowed to attend, and the admission of a few ill-favoured applicants made the schoolroom a fair sample of local childhood, with the result that in examination a less favourable report was made than would be the case if the best children only were admitted.

These points are mentioned not as praise or in criticism, but to illustrate that a correct allowance for the material in a school is fair in estimating the money grant earned, and that the care of the feeble children, as well as of the average, may be encouraged by judicious assessment of the child-material and their intellectual culture.

On the other hand certain philanthropic institutions claim credit for the badness of the material collected and justly claim support on the ground that it is to the public interest that the best should be done for the paralysed, the crippled, the blind, &c., and for those of criminal tendencies. If these principles be accepted, if it is true that the national interest lies in having a population of children well made and well trained, is it to our advantage that the day school system should ignore or reject the weak and defective children, and might they not in appropriate cases be kept under their parents' care, and at day schools, with due allowance to the teachers, thus being retained in their own neighbourhood?

Certain schools, if not all, and in particular the endowed schools of a parish should be encouraged to show results in dealing with the worst child material of the neighbourhood, and in preventing it from gravitating to the Poor Law, or to degradation and failure.

Assessment of Results of Intellectual and Physical Training in a School, making allowance for the Physical Condition of the Children.

Having obtained a report on the physical condition of the children in a school we may proceed to estimate the number of dull children and the number of nerve cases to be expected making allowance for the material in the school. The estimate is founded upon the conditions seen and in comparison with a now established average of 100,000 children.

There are two results of physical training characterising it as satisfactory; (1) in cases of defective development to remove or prevent abnormal nerve-signs; (2) in children of normal development to prevent, or at least not to produce nerve-signs. Thus taking the development cases in a school, a high percentage of abnormal nerve-signs among them is against the effects of the training, and a high percentage of nerve cases without defects in development suggests that the training is not good. So also in each case a high percentage of mental dulness shows want of adaptation of methods of teaching to the special requirments of the children.

We may estimate as Dull Children (seen 1888-91).

	Boys %	Girls %
Of development cases with nerve-signs ..	43	43
Of development cases without nerve-signs .	33	39
Of nerve cases without development defects	37	41

We may estimate as the average number of cases with Nerve-signs.

	Boys %	Girls %
Of the development cases	55	49
Of the total number of children seen as presenting defects in development	5	4

Educationalists may desire a further knowledge of the conditions of child life, and the groups of children that have to be cared for, as an aid to the solution of many points in school management.

What modifications of the ordinary course of education are needed for—

1. Cases feeble-minded, or semi-imbecile, those admissible as candidates for an asylum, but also suitable for admission to classes of special instruction in day schools.
2. Children feebly-gifted, motionless, statuesque or dull.
3. Delicate but bright children.
4. Children mentally bright but defective in moral sense.

5. Children well made, but exhausted temporarily, or as a chronic condition.
6. Children partially deaf.
7. Children crippled or paralysed.
8. Epileptics, specially such as are harmless in a day school.
9. Eye cases.

What are the conditions of children and the consequent educational requirements in the following group?

10. The poor class.
11. The wealthy class.
12. High-pressure schools.
13. Schools considered "inefficient."

The Training Colleges.

It is to these we must look for the trained teachers; and to University provision for their further training. It is possible to systematise this kind of work, and to give instruction to candidates for the teaching profession without touching upon medicine, such as may enable them to observe for themselves and make deductions from what they see; such training should be based upon scientific knowledge, and the methods of mental-action may be taught by reference to the visible facts of its expression and its application to the study of psychology and the conditions of child-life. Many subjects of importance to teachers might be dwelt upon, but we shall only give a few illustrative points. When the children are allowed to choose their seats in school, something like a process of natural selection occurs; nervous children are gregarious, defective children are solitary; the former are usually bright at lessons, and congregate on the back seats, where they often do their work and then play, while the duller children are kept in front under the teacher's eye. Eye-movements need training, and where this is neglected in the infants' school, results in learning to read are apt to be slow, and some children are bad and inaccurate observers.

The speech of all children needs careful cultivation, and incipient stammering will best be combated by the teacher who observes the first indications of the spasm affecting the face.

Imitation as an Element in Training.

This is mainly exercised through sight, and is universally employed by teachers, specially among infants. When the pupil faces the teacher his right hand is opposite to the teacher's left. Experience shows that when told to do as the teacher does, if the teacher raises his right hand the pupil tends to hold out his left. So strongly is this faculty marked that, to obtain results in action, many teachers habitually use their left hand for imitation by the child's right. It seems hardly surprising then that children confuse their right hand with the left when subsequently directed by word of command. Imitation by vision is probably one of the most direct means by which the teacher knowingly or unconsciously controls the pupil's brain—a means of transference of nerve-status often called a moral influence, still a very real, true, and powerful one for good or ill.

Military Drill.

Boys thus trained present much uniformity of action, and response to verbal command is quick. When examined in groups these good effects are obvious when examining a boy individually, specially if the master be not at hand, the finer movements and symmetry of balance is usually lost, and it often seemed that less true balance and self-command over the nerve-system was present than among children trained by free exercises.

Each method of physical training has its advantages, the effects must be observed in the individual child alone.

GENERAL RESULTS.

The work that has already been done affords a considerable amount of evidence to the following propositions :—

It is practicable to inspect, report upon, and classify the children seen in a school by means of facts seen and the teacher's report. Evidence of scientific value is thus obtainable of importance to the State, to education, and to philanthropic efforts.

The average child material in a school or district may be determined. The conditions of development and the nerve-signs vary very much in different schools ; as to the latter, observation suggests that adapted methods of training may remove them.

The co-relation of visible signs with low nutrition and mental dulness has in many cases been demonstrated.

Ill-made and feeble children tend to gravitate to the Poor Law and Certified Industrial Schools, and to the lower standards of day schools. The want of provision for mentally feeble children in day schools, and in many cases their exemption on medical certificates, tends to throw such cases upon the care of the State, and many become degraded.

Feebly-gifted children, the paralysed, and in some cases the epileptic, may in many cases and in limited numbers be educated in special classes in day schools.

If any special knowledge concerning the condition of the child-population in our towns and sanitary areas is to be obtained, and if we desire to remove the defects among them and improve the average standard, then   is highly desirable that the modes of procedure should be systematized, and plans arranged for carrying on such work as has been explained, as a Department of State Medicine, side by side and in aid of our Public Hygiene.

Having given some account of the 50,000 children seen 1888—91, we may proceed to arrange them in groups, as seen in resident schools and day schools, dividing the latter into sub-groups of higher and lower social standing; lastly the children may be examined as seen in schools for English, Irish, and Jew children.

The facts referred to are given in Tables XIX. and XX., and we may take as a basis for comparison the average for the 50,000 cases; the tones of figures representing conditions may be higher or lower than this average.

The percentages go against the resident schools, except as to bodily nutrition. The material they receive is worse than the average, and their results are not so good; this is accentuated in the certified industrial schools, and less marked in the homes and orphanages.

Taking the average condition of the day schools and comparing with it the children of upper grade with those of poorer class, there is a preponderance in favour of the lower class as less dull or defective.

As to the nationalities the differences are more marked than among the social classes. In day schools the Jewish children—the families of the Whitechapel Jew immigrants—stand as by far the best in all conditions: Development cases 7·5 per cent., Dull 5·1 per cent., Low Nutrition 2·7 per cent.; as against the English children with Development cases 10·8 per cent., Dull 6·8 per cent., Low Nutrition 4·3 per cent.; and in all Irish with Development cases 20·0 per cent., Dull 13 per cent., Low Nutrition 5·5 per cent.

School Organisation.

School organisation by the teachers is mainly founded upon their experience of the child's mental ability and work in school. This takes time and frequently a new pupil is not placed in a suitable standard till some weeks' experience shows the child's mental capacity. A knowledge of the points observed in this enquiry might greatly facilitate the responsible work of classification for educational purposes. Two Standards frequently, though not always met with in schools, call for special remark. In Standard O or Primers the children are collected who, being over age for the Infant School, are still too backward for Standard I. In Standard Ex VII. we find the children who have passed through the ordinary classes of the school. Among the 63 schools seen, 1892-94, there were 25 · with either Standard Ex VII. or Standard O at the visit. The numbers seen in these standards is indicated in the following Table.

TABLE VIII.—*Cases seen* 1892—94.

	Total number seen in schools.		Ex VII. only.		Stand'rd O only.		Ex VII. & Stand'rd O.			
							Ex VII.		Stand'rd O	
	Boys.	Girls.	B.	G.	B.	G.	B.	G.	B.	G.
11 schools with Standard Ex VII. only............	6,373	5,727	110	101	...	...	...	...	...	...
10 schools with Standard O only	4,508	4,143	...	...	255	212	...	...	...	...
4 schools with Standard Ex VII. & Standard O	2,362	2,107	...	...	...	...	34	30	99	111
25	13,243	11,977	110	101	255	212	34	30	99	111

In these 25 schools we find as follows :—

In Standard Ex VII.: Boys, 144; Girls, 131. Reported as dull: Boys, 4; Girls, 5. In Standard O: Boys, 354; Girls, 323. Reported as dull: Boys, 93; Girls, 107. For conditions of children in these Standards see Table XXI.

More accumulations of dull children in a certain class, whether a class of Primers, or in a lower section of Standard III. for older children, may make the other class rooms brighter; but when children below the average in mental power are accumulated, there arises a greater responsibility for their individual care, which must be met by the provision of a sufficient staff of specially-trained teachers.

Secondary education, such as is carried on in the Standard Ex VII. of our public elementary schools, tends to accentuate the difficulties arising from the classification of children solely according to mental status. In elementary schools of higher grade, a boy entering Standard I. in the boys' school is unacceptable unless he can work well; after a certain age, the dull boy cannot conveniently be kept in the infant school, for which he is too big. He must then either be kept among the infants, for whom he is not good company, or go among classmates with whom he cannot profitably work. To meet such cases it often happens that there is a class of Primers or Standard O, but without any special arrangements for individual culture of these dull or backward pupils.

In such schools the brighter children are well educated; at fourteen years of age they get the prizes of the school and enter social life at an advantage; the dull children on the other hand have not only been left comparatively uncultured, but by raising a class distinctly superior to them, they find the struggle for existence becoming intensified.

" Schools of Special Difficulty."

Until the year 1890, certain schools under the London School Board were officially recognized as schools of special difficulty; three of these schools were visited in 1892 and the fourth in 1894. An account of the children in these schools is given in Tables XXVI., XXVII. and XXVIII. On reference to Table XXVIII. it will be seen that while presenting a percentage of dull pupils much higher than the average of Board Schools, they also contained more than the average number of children with all the Main Classes of defects.

Schools where the Number of Children in Attendance is Small in Relation to the School Accommodation.

In two London Board Schools, with an aggregate accommodation of 1828 places, there were present only 566 boys and 451 girls: this was explained by the diminishing child-population of the neighbourhoods, and both schools were about to be closed. The conditions of these children are indicated in Tables XXVI., XXVII. and XXVIII., and on reference to Table XXVIII., it will be seen that the child-material was much below the average and presented a super-abundance of abnormal nerve-signs.

CHAPTER VIII.

RECOMMENDATIONS.

A—Appointment of a Scientific Commission of Enquiry by Government.

It is desirable that a small Scientific Commission of Enquiry should be appointed by the Government for the purpose of determining the conditions of portions of the School population as to their mental and physical power; ascertaining the numbers of such as are of imperfect development, their distribution, and the possible causes of such defects.

B—Need of Enquiry as to Feeble-minded Children.

In view of the harm resulting not only to the individual, but also to society from the educational neglect of the feeble-minded child (the defective child growing up dependent, and possibly delinquent) it is expedient that the State should officially obtain information as to the social necessities of this class, and of the physical signs of mental deficiency indicating the need for special training. Investigations with this object should be made in various urban and rural districts throughout the country, and may incidentally furnish information as to the influence of the special circumstances and occupations of a given locality upon the mental and physical conditions of the population.

C—Desirability of a Parliamentary Enquiry as to Feeble-minded Children, their Condition and Training.

It is desirable that a Committee be appointed by the House of Lords, or other public body, to examine the evidence now afforded as to the conditions of childhood; the best means of dealing with children who are dull or deficient; the means at present available for their educational care; and to report generally upon the status and needs of dull and defective children, and the working of methods of education as they affect such children.

E

D—Recommendations of the Royal Commission on Blind, Dumb, &c., as to the Feeble-minded.

The Royal Commission on the Blind and Dumb, &c., having reported — " That with regard to ' feeble-minded ' children, they should be separated from ordinary scholars in public elementary schools, in order that they may receive special instruction; and that the attention of school authorities be particularly directed to this object." It is recommended that the Act to make better provision for the Elementary Education of Blind and Deaf Children in England and Wales [56 and 57 Vict., chap. 42], should be extended to include children with other mental and bodily defects as well as the epileptic.

E—Expert Scientific Advice to Government Departments.

That the Government be recommended to appoint scientific experts to assist the Education Department, the Local Government Board, and the Home Office, with regard to the means to be adopted for the education of children requiring special care.

CHAPTER IX.

SUGGESTIONS AS TO THE CONDUCT AND ADMINISTRATION OF THE EDUCATION OF FEEBLE-MINDED CHILDREN.

RECOMMENDATIONS AS TO ADMINISTRATION.

1—School Board Census should note Mentally-feeble and Afflicted Children.

That School Boards, in taking the triennial census of their district, should register any mentally-defective children, or children otherwise afflicted.

2—Feeble-minded Children should attend a Special School or Class.

The fact that a child is found to be feeble-minded is no reason for excluding him from school attendance, but special arrangements should be made by school authorities to provide such teaching as is appropriate to feeble-minded children. These cases appear to be not more than one or one and a half per cent. of the children seen.

3—Feeble-minded Children to be trained in Day Schools or Boarded Institutions.

In making provision for the special instruction and training of "feeble-minded" children, the character of the home surroundings and the care which the child will receive out of school should be taken into consideration in determining whether attendance at a day school or residence in a boarding institution is preferable.

4—Day School Classes of Special Instruction.

It appears desirable that day schools or classes of special instruction should be established at a sufficient number of centres for "feeble-minded " children. This is needed primarily that such children may have special instruction provided for them and not be excluded from school attendance ; pupils from other schools might be conveniently transferred to them so as to meet local requirements.

5—*Report on Individual Feeble-minded Children in Schools.*

It is desirable that children in schools or classes for "the feeble-minded" should be separately and individually reported on as to mental and physical condition, both on admission to the school, and after definite periods of instruction; and that for this purpose independent Government Inspectors should be appointed for the Public Elementary, Poor Law, and Industrial schools.

6—*Trained Teaching Staff for Feeble-minded Children.*

That, in order to provide a staff to undertake the educational care of weak and mentally-feeble children, a special course of training for such teachers should be arranged.

7—*Lectures and Training for Teachers of the Feeble-minded.*

That lectures should be instituted on the observation, study, and classification of children, as to conditions bearing on mental life and education. This might consist of an elementary course and of University teaching. Arrangements should be included for demonstrations in a practising school.

8—*Discrimination and Report on Feeble-minded and Afflicted Children.*

The selection of and report on children who are feeble-minded or who on other grounds require special care and training, should be made upon a methodical plan by a Medical Officer. Children found to be absent from school on account of mental weakness or bodily affliction, as well as pupils presented by teachers as unfitted for the ordinary classes should be carefully reported on in order that suitable training may be provided for classified groups of children.

9—Certificate as to a child requiring Special Educational Training.

Name of child.
Age.
Address.
Physical health and condition.
Developmental defects.
Nervous defects.
Defects in mental power.
Facts communicated by others (stating from whom).
Opinion and recommendation as to the case.

Date.   (Signed).   Medical Officer.

10—Classes for Dull and Backward Children not Feeble-minded.

Where the school organization includes a special class for the dull and backward children as with Primers or Standard O, which mostly contains the children too old for the Infant School and too backward for Standard I. ; or a class where older children who are dull are accumulated higher up in the school, the arrangements, the number of pupils in each class, and the selection of the teacher, should be adapted to the special difficulties of dealing with these children whom it has been shown particularly need careful and individual attention.

11—Instruction for Teachers in Training Colleges.

Special practical and theoretical instruction should be given to all teachers in Training Colleges and elsewhere as to the physical indications of weakness and mental-feebleness in children, and as to the points to which they should direct attention in school classification and teaching—thus enabling them to describe and if necessary report on individual children to the School Authority.

12—The Special Care of Dull and Feeble-minded Children is imperative.

The educational care of Dull children and those Feebly-gifted mentally, is as much a duty devolving on the Community as is the Education of bright and healthy children.

13—Children in Punitive Schools under the Home Office should be individually reported on.

In certified industrial and other schools under the Home Office, where children are detained for lengthened periods under a magistrate's order, it is of special importance, with regard to preventing mistaken harshness, and in consideration of the punitive discipline of such establishments, that a report should be made upon the admission of an inmate as to his mental and physical condition, in addition to an annual report upon each child, in order that cases of bodily and brain defect may be dealt with according to circumstances. It has been shown that cases of defectiveness are very common in such establishments.

14—Recommendation to Boards of Guardians relative to Feeble-minded Children.

That the attention of Boards of Guardians be specially directed to the importance of discriminating the dull, backward, and mentally-deficient children under their care, and to the best means of providing for their special training, and to the desirability of preparing a special annual report concerning the mentally-deficient and otherwise afflicted children for whom they are responsible.

15—Feeble-minded persons in Workhouses and those seeking relief.

That Guardians should prepare returns showing the number of " weak-minded " applicants for relief, and inmates of workhouses with whom they have to deal.

16—Recommendation to Philanthropic Societies as to discrimination of the " Feeble-minded."

That the attention of institutions, societies and philanthropic bodies dealing with dependent and afflicted children and young persons, be directed to the careful discrimination of the weak-minded for whom special provision is necessary.

This is specially important in dealing with young people crippled, paralysed, epileptic, and those with a tendency to delinquency as well as with others failing in self-dependence.

All such bodies are strongly recommended to seek professional assistance and advice, both when undertaking the care of cases and in dealing with special difficulties as they rise.

17—*Preparation of Vital Statistics.*

That in the preparation of all vital statistics males should be distinguished from females, and that in returns of mortality Developmental diseases should always be included.

DOUGLAS GALTON,
Chairman of Council.

E. WHITE WALLIS
Secretary.

June, 1895.

NOMENCLATURE OF SIGNS ADOPTED AS THE BASIS OF THIS ENQUIRY.—WITH TABULATION OF CASES.

A further account of the signs or abnormal points observed, is given in Chap. IV., page 20.

The columns indicate the numbers of children presenting each condition or defect respectively as seen in school, 1888—91 and 1892—94. The total numbers among the 100,000 children under report are also given. If a decimal point be inserted at the third numeral from the right hand it gives the percentage of the number of children with the condition or defect as taken upon the number of children seen.

Each defect is enumerated, whether alone or in combination with other defects.	Of 50,000 in Schools seen 1888—1891.		Of 50,000 in Schools seen 1892—1894.		Of total of 100,000 in Schools seen 1888—1894.
	Boys.	Girls.	Boys.	Girls.	Boys & Girls.
S. Number of children seen ...	26,857*	23,143	26,287	23,713	100,000
A. DEFECTS IN DEVELOPMENT.—The term includes any point of defect in the form, proportion, or size of the body and its parts, or the absence of any part ...	3,616	2,235	2,308	1,618	9,777
(a 1). CRANIUM Defective includes any defect in size, form, proportions, or ossification of the cranium. A given case may come under more than one of the classes below. As to a standard of normal size : In a well-developed child of good potentiality the head circumference at 9th month is 17½ inches, at 12 months 19 inches, at 7 years 20 to 21 inches. This is a rather high standard of size ...	1,528	1,048	806	611	3,993
(a 2). CRANIUM Large.—A head of 22 inches circumference or over may be considered large in a school child; allowance must be made for age. Doubtless many of these cases are rachitic. Hydrocephalus is entered in its own class ...	257	46	107	13	423

(a 3). CRANIUM SMALL.—The point of size of head is recorded as apart from the size of the child for its age. The volume is estimated in relation to the normal for age. This is determined by inspection, by the open hand placed upon the head, and by the measuring tape. A head with circumference over 20 inches at any school age is not registered as small; usually the small heads are 18 to 19¾ inches circumference. Small head is noted independent of stature ... ...

(a 4). CRANIUM BOSSED.—There may be bosses, protuberances, or outgrowth at the sites of the ossific centres of the frontal bones, at the parietal centres, at the site of the fontanelle, and elsewhere. These are usually symmetrical, but not always ... ...

(a 5). Forehead Defective.—The forehead may be narrow, shallow in vertical measurement, or small in all dimensions; it may bulge forward and over-hang. All defects of the forehead, except "bosses" and "frontal ridge" (a 4, a 6) are here included ... ...

(a 6). Interfrontal Ridge.—The vertical suture between the two halves of the frontal bone may be the site of a bony ridge, present in all degrees; if the forehead be also narrow it forms the scapho-cephalic type ...

(a 7). CRANIUM Asymmetrical.—Asymmetry may be as to the forehead or other part; one side of the cranium may be smaller than the other ...

(a 8). Dolichocephalic.—Head long in antero-posterior diameter ...

(a 9). Hydrocephalic.—This term is used as in medicine ... ...

(a 10). Other types of CRANIUM.—Square; oxycephalic, or elevated and conical; cranium larger in anterior than in posterior segment... ...

(b 11). EXTERNAL EAR Defective in its parts, size, or form. Abnormality in size, proportioning, absence of parts, texture of skin are here recorded. The ear may be outstanding with great convexity posteriorly and concavity in front; the helix or portions thereof and the antehelix may be absent; the skin over the cartilage may be tight and adherent, coarse in texture with varicosities. The ears may be asymmetrical, and the lobes may be adherent to the face ... ...

Description					
(a 3). Cranium Small	327	738	149	516	1,730
(a 4). Cranium Bossed	495	127	323	47	992
(a 5). Forehead Defective	183	78	53	23	337
(a 6). Interfrontal Ridge	89	27	121	19	256
(a 7). Cranium Asymmetrical	84	16	27	2	129
(a 8). Dolichocephalic	43	10	26	2	81
(a 9). Hydrocephalic	5	2	2	1	10
(a 10). Other types of Cranium	50	6	11	2	69
(b 11). External Ear Defective	1,047	268	304	103	1,782

The actual number seen was 26,884, but 27 have been deducted so as to make a total of 50,000 for convenience of working percentages.

Each defect is enumerated, whether alone or in combination with other defects.	Of 50,000 in Schools seen 1888—1891.		Of 50,000 in Schools seen 1892—1894.		Of total of 100,000 in Schools seen 1888—1894.
	Boys.	Girls.	Boys.	Girls.	Boys & Girls
(c 12). EYELIDS WITH EPICANTHIS.—The epicanthis is a fold of skin continuous with the lower fold of the upper eyelid (not a fold of mucous membrane) placed across the inner angle of the opening of the eyelids covering the caruncle; it may be asymmetrical ...	514	384	288	190	1,376
(d 13). PALATE Defective in Shape.—Defects in form are described as seen in the horizontal and in the vertical plane ...	796	525	496	310	2,127
(d 14). PALATE Narrow.—Without being otherwise altered, the palate may be contracted laterally in the space between the alveolar processes ...	450	291	276	163	1,180
(d 15). V-shaped PALATE.—Pointed more or less sharply at its anterior extremity, the alveolar processes being nearly straight lines, meeting at their extremities at an acute angle ...	235	171	179	110	605
(d 16). PALATE Arched or Vaulted, thus deviating from the normal in the vertical plane with a high roof ...	86	41	30	22	179
(d 17). PALATE Cleft.—A deformity which may affect the hard and the soft palate ...	14	8	12	13	47
(d 18). Other Defective Types of PALATE, such as the flat and the horse-shoe type ...	10	15	......	3	28
(e 19). NASAL BONES, wide, sunken, or indented. The bony bridge of the nose may be thus ill-shapen and depressed as in the undeveloped condition of babyhood ...	241	214	155	153	763
(f 20). GROWTH SMALL or stature short. Children short and small in build for their age ...	209	209	271	328	1,017
(g 21). OTHER DEFECTS IN DEVELOPMENT less frequently observed ...	908	645	250	212	2,015
(g 22). Adipose Type.—Children fat and flabby, generally without spontaneity and slow in action ...	16	9	1	4	30

Description					
(g 23). Cyanosis.—General blueness of face, lips, and ears, commonly dependent on defect of the heart	2	......	2	......	......
(g 24). Dermoid Cyst.—Tumours at margin of orbit or near the temporal fossa	......	......	......	......	7
(g 25). Face Asymmetrical, one side being smaller than the other	13	1	3	2	18
(g 26). Face Small.—The face, including the upper and lower jaws, with their bones, may be small, independent of the size of the calvarium or brain case of the skull	67	21	17	11	147
(g 27). Features Coarse, heavy, flat, or lips thick. The features may be large and ill-proportioned. The separate features may not be individually malformed, but disproportionate one to another or to the size of the face; thus the nose may be small, the face large, round, flat, the features rising from the plane of the face. The lips may be thick and protuberant	290	12	27	104	40
(g 28). Forehead Hairy.—The forehead may be covered with downy hair; the hairy scalp may join the outer extremities of the eyebrows	66	4	19	3	13
(g 29). Frontal Veins Large.—There may be well-marked veins in the middle of the forehead and across the bridge of the nose	35	4	12	6	17
(g 30). Hands Blue and Cold.—This was registered when it appeared to be a more or less permanent condition as a defect independent of weather. It may be seen in a paralysed limb	88	47	13	11	10
(g 31). Hare Lip.—Congenital fissure of the upper lip	37	7	15	5	10
(g 32). Ichthyosis.—Scaly skin on wrists and arms, or general on face, ears, and all parts	24	2	5	7	1
(g 33). Moles on face or eyebrows; they may be pigmented, and may or may not be raised, and are often covered with hair	9	6	1	1	27
(g 34). Mouth Small.—Referring to measurement of the opening when the face is at rest	132	38	50	17	3
(g 35). Nœvus.—"Port wine stains" or patches on the face coloured by vascularity. Half the face may be thus affected, with affection of conjunctiva and mucous membrane of mouth	12	2	4	3	33
(g 36). Nose Soft Tissue, wide or superabundant. The skin and subcutaneous tissue of the bridge of nose may be superabundant and wide, giving an appearance of great width between the eyes	91	17	13	28	

Each defect is enumerated, whether alone or in combination with other defects.	Of 50,000 in Schools seen 1888—1891.		Of 50,000 in Schools seen 1892—1894.		Of total of 100,000 in Schools seen 1888—1894.
	Boys.	Girls.	Boys.	Girls.	Boys & Girls.
(g 37). Orbits Oblique.—The transverse axis of the orbits sloping in place of being horizontal...	3	5	......	......	8
(g 38). Orbits Sunken.—The whole cavity and its malar boundary appearing sunken into the skull ...	3	2	......	......	5
(g 39). Palpebral Fissures Defective in size or form. The eyelids may be small as well as the palpebral fissures or openings between them, both in their vertical and transverse measurements. In some cases the opening is not symmetrical, being wider on the inner than on its outer half. The transverse axis may slope outwards and upwards, or outwards and downwards, instead of being horizontal ...	98	83	41	25	247
(g 40). Prognathous Type.—The lower jaw large, heavy, underhung ...	22	......	......	......	22
(g 41). Supernumerary Ears represented by sessile or pedunculated outgrowths in front of the tragus, sometimes nearly half an inch long ...	7	3	21	12	43
(g 42). Miscellaneous Defects in Development.—Under this heading are included congenital defects of eyes and congenital deformities of the body ...	......	......	12	15	27
B. ABNORMAL NERVE SIGNS seen in the balances and movements of the body ...	3,413	2,074	2,853	2,015	10,355
(43). General Balance Defective.—Asymmetrical positions of the limbs, shoulders, back; slouching, listless gait ...	201	173	90	113	577
(44). Expression Defective.—Want of changefulness, vacancy, fixed expression. The visible muscular action and balance seen in a face may be described, and still there may be an expression that cannot be described anatomically. A face may be balanced or moved abnormally by action of its muscles, and yet carry a good expression ... (45). Frontals Overacting.—The frontal muscles may produce horizontal creases in the forehead, which may be deep if these muscles overact coarsely.	694	474	151	191	1,510

Description					Total
Sometimes these muscles are seen working under the skin in vermicular fashion, with an athetoid movement; in other cases the action is fine, producing what may be called a dull forehead. This over action does not necessarily erase expression ...	1,322	294	696	146	2,458
(46). Corrugation.—Knitting the eyebrows, drawing the eyebrows together; vertical creases are thus produced on the forehead above the nose. Corrugation may co-exist with over-action in the frontal muscles ...	199	40	38	11	288
(47). Orbicularis Oculi Relaxed.—There is a thin muscle, the orbicularis oculi, which encircles the eyelids. Its tone gives sharpness to the lower lid, so that its convexity is seen. When this muscle is relaxed there is a fulness or bagginess under the eyes, which is not due to œdema (dropsy), and may disappear on laughter ...	522	343	371	293	1,529
(48). Eye Movements Defective.—When an object is moved at a distance two feet in front of the face, the eyes normally move in following it; in some children the head always turns towards the object, while the eyes are kept still in their orbits. In other cases fixation of the eyes is bad, or there are restless, uncontrolled movements of the eyes ...	798	485	348	261	1,892
(49). Head Balance Asymmetrical or Drooped.—In the normal the head is held erect. It may be inclined to one side or drooped ...	219	319	95	274	907
(50). Hand Balance Weak.—In this type of balance the wrist is slightly dropped, the palm is contracted laterally, and the digits are slightly bent ...	715	504	1,234	778	3,231
(51). Hand Balance Nervous.—When the arms are held out the wrist droops, the palm is slightly contracted laterally, the thumb and fingers are extended backwards beyond the straight line at their junction with the palm ...	550	516	253	359	1,078
(52). Finger Twitches.—When the hands are held out for inspection, there may be twitching movements of the digits in flexion, or extension, or laterally ...	445	261	145	142	993
(53). Lordosis.—When the hands are held forward, an alteration in the balance of the spine may appear, with an arching forward in the lower part of the back, while the upper part of the spine between the shoulders is thrown back ...	185	279	36	112	612

78

Each defect is enumerated, whether alone or in combination with other defects.	Of 50,000 in Schools seen 1888—1891.		Of 50,000 in Schools seen 1892—1894.		Of total of 100,000 in Schools seen 1888—1894.
	Boys.	Girls.	Boys.	Girls.	Boys & Girls.
(h 51). OTHER ABNORMAL NERVE SIGNS less frequently observed. Signs grouped for convenience of primary arrangement as being less frequent in occurrence than those given earlier, but not necessarily of less importance ...					
(55). Deaf, or Hearing Defective.—Children deaf, or partially deaf. Tests for hearing cannot be used in a school enquiry ...	434	234	468	282	1,418
(56). Grinning, or Over-smiling.—Over-smiling or grinning may be spontaneous, or may occur on any stimulation to effort. The lines formed in the naso-labial region of the face may be fine or coarse; there may be a duplicate or triplicate naso-labial groove partly depending upon the thinness or thickness of the skin; permanent skin creases may result ...	34	33	12	15	94
(57). Mouth Open or Jaw Drooped.—The jaw may be drooped, or the mouth may be open with the teeth closed. This should not be recorded as a nerve sign if it is probably due to obstruction of the respiratory passages ...	69	43	19	14	145
(58). Over-Mobile.—Constant spontaneous movements. Among children in the infant school, and in some over 7 years, spontaneous movement is normal; it is most common in the fingers ...	134	59	233	110	536
(59). Response in Action Defective.—Response in action following a command or in imitation may be accurate or uncertain, prompt or slow. There may be an interval between the command and the response, or the action may be continued unduly long. Response may be better when stimulated through the eye or through the ear respectively ...	4	......	4	9	17
(60). Speech Defective.—Stammering (spasm), or defect in articulating certain sounds. Speech may be nearly absent; it may be indistinct. As a mental defect, the question asked may be repeated without a reply ...	112	56	56	62	286
	116	70	105	52	343

2	5	4	......	11
30	11	15	14	70
434	234	67	45	780
1,030	973	749	770	3,522
2,216	1,463	2,077	1,635	7,391
836	637	764	692	2,929
485	322	470	345	1,622
124	152	142	226	644
25	23	39	55	142
8	11	12	11	42
75	54	52	46	227

(61). Statuesque or Immobile.—Without any spontaneity, immobile except under stimulation to action ... ...

(62). Tremor.—A uniform rapid movement with but slight displacement of the parts moving ... ...

(63). Miscellaneous Abnormal Nerve Signs.—Under this heading are included eye cases (71), (72), and the paralytic cripples ... ...

C. Nutrition Low, Thin, Pale, Signs of Delicacy.—This was registered to any child seen to be pale, thin, or delicate. No enquiries were made as to the feeding of the children. Nutrition of the limbs and face was observed, as well as colour in the face and lips ... ...

D. Dull Mentally, or so Reported by the Teachers.—In every case registered the teacher's opinion concerning the child's mental capacity was asked and written down; those reported as below average ability in school were registered as dull. After the children presenting visible defects had been picked out, the teachers were invited to present any other pupils known by them to be mentally dull. All grades of mental dulness were registered under this heading—see (76), (77), (78) ... ...

E. Defects of Eyes.—When the eyes were looked at obvious defects were noted, but no tests were used as to acuteness of vision or errors of refraction, and the ophthalmoscope could not be used in the schools. Ophthalmia was not registered, but some of its late effects are recorded under "Disease of Cornea," "Eye lost by Disease" (68), (70) ... ...

(64). Squint.—Under this heading are registered cases of organic squint, one eye being turned; also temporary or varying convergence when looking at an object two feet from the face, which probably indicates hypermetropia ..

(65). Using Convex Glasses.—Evidence of hypermetropia or long sight ...

(66). Using Concave Glasses.—Evidence of myopia or short sight ... ...

(67). Myopia not Using Glasses.—Short sight ascertained on inquiry ...

(68). Disease of Cornea.—Inflammation, ulcers, white patches. It was not found convenient to record ophthalmia, but if disease of cornea were present this was registered ... ...

80

Each defect is enumerated, whether alone or in combination with other defects.	Of 50,000 in Schools seen 1888—1891		Of 50,000 in Schools seen 1892—1894		Of total of 100,000 in Schools seen 1888—1894
	Boys.	Girls.	Boys.	Girls.	Boys & Girls.
(69). Eye Lost by Accident.—As stated on inquiry	18	18	33	16	85
(70). Eye Lost by Disease.—Inquiry was made as to cause of loss of eye	18	12	10	18	58
(71). Nystagmus.—Organic tremor of the eyes. This defect is also registered under "Other Nerve Signs. Miscellaneous. (63)"	25	9	20	11	65
(72). Ptosis.—Drooping of eyelid may be partial or complete in one eye or in both. This defect is also registered under "Other Nerve Signs. Miscellaneous. (63)"	26	11	24	5	66
(73). Pupils unequal.—Inequality of size of pupils when the eyes are equally stimulated by light	2		4	2	11
(74). Cataract may be congenital or the result of injury		3	8	5	13
(75). Miscellaneous and Congenital Defects of the Eyes.—Including coloboma or defect in a portion of the iris. Unequal and asymmetrical pigmentation of irides and "tortoise-shell irides." Albinos destitute of all pigmentation. Excessive largeness of cornea. Congenital smallness of one eye. Congenital blindness from cause unknown. Congenital defects of the eyes are also registered under "Other Defects of Development. Miscellaneous. (42)"	20	16	8	8	52
F. Cases of Rickets.—When bent legs or pigeon breast indicated rickets the fact was registered; conditions of the cranium were not accepted as evidence of rickets, but were registered under their respective headings. (See Aa 2, 4, 5.)	157	39	39	9	244
G. Exceptional Children.—Children who on account of certain observed defects may at once be said to need individual consideration	303	204	157	147	811
(76). Idiots. (See Catalogue, Group 6.)	2	......	......	......	2
(77). Imbeciles. (See Catalogue, Group 7.)	30	16	3	2	51
(78). Children feebly gifted mentally. (See Catalogue, Group 8.)	89	85	49	52	275
(79). Children mentally exceptional. (See Catalogue, Group 9.)	3	9	4	3	19

					Total
(80). Epileptics, and children with history of fits during school life. (See Catalogue, Group 10.)	36	18	21	35	110
(81). Dumb			4	1	5
(‡82). CHILDREN CRIPPLED, maimed, and paralysed. (See Catalogue, Group 11.)					
(83). Disease of hip	155	84	75	60	374
(84). Disease of spine	24	18	11	3	56
(85). Disease of upper limb	25	17	11	8	61
(86). Disease of lower limb	11	2	7	3	23
(87). Hand maimed	13	7	11	5	36
(88). Amputation of arm			2	1	3
(89). Amputation of leg		1			1
(90). Congenital absence of greater part of upper limb	15	8	5	1	29
(91). Congenital absence of hand	3	2	2	3	10
(92). Congenital defect of hand		4			4
(93). Congenital absence of foot	4	2		2	8
(94). Club foot		1	1		2
(95). Hemiplegia	14	9	7	11	41
(96). Paraplegia	2	2			4
(97). Infantile palsy, upper limb	10	4	1	3	18
(98). Infantile palsy, lower limb	32	6	11	10	59
(99). Torticollis	1	1	4	5	11
(100). Blind, or nearly so				3	3
(101). Chorea				1	1
(102). Crippled by burn			2	4	6
(103). Heart disease			1		1
(104). Facial paralysis	1				1

CATALOGUE OF GROUPS OF CHILDREN AND GROUPS OF CASES.

An essential principle in this Report is to describe children in terms indicating Physical conditions ; so that it may be possible in further discussion as to the Management and Training of children to express with some accuracy the kind of children under special consideration, and to ascertain from the Tables in the Report the Distribution and Co-relations of such cases as are found among the samples of child-life here examined.

Throughout the text and Tables the various groups are for brevity denoted by the alphabetical symbols allotted to them in the Catalogue. The Primary groups are denoted by capital letters, the Compound groups by capital letters followed by the sign +. S is used for children seen. N for children noted.

The columns indicate the numbers of children in the groups respectively as seen in schools, 1888-91, and 1892-94. The total numbers among the 100,000 children under report are also given ; if a decimal point be inserted at the third numeral we have the percentage of the number of children in the group as taken upon the number of children seen. A further account of these Groups of Cases is given in Chap. VI., page 38.

The first twelve groups of children are arranged according to conditions of obvious importance, and are of special, social, and educational interest.

Group.	Symbol.		Number of Children in which some defect was noted.				
			Of 50,000 seen in 1888—1891.		Of 50,000 seen in 1892—1894.		Of total 100,000 1888-94.
			Boys.	Girls.	Boys.	Girls.	B. & G.
	S	Total number of children seen	26,857*	23,143	26,287	23,713	100,000
1.	S—N	NORMAL CHILDREN.—This includes all children not presenting any visible defect in Development, Nutrition or physical condition, with no abnormal Nerve-signs, and not reported as Mentally dull. The numbers in the school Standards respectively are recorded for enquiry, 1892-94	21,305	19,536	21,175	19,884	81,900
2.	N	CHILDREN NOTED AND REGISTERED.—This includes all children presenting any of the signs or defects given in the Nomenclature of defects. Each of these children presented one or more of these defects or was Mentally dull ; a schedule was filled in for each of these cases ...	5,579	3,607	5,112	3,829	18,127

3.	E	EYE CASES.—This includes all children showing any of the defects of eyes enumerated in the Nomenclature of defects (64 to 75). Tests for acuteness of vision or errors of refraction could not be used in the schools, but when spectacles were used, or when inspection showed obvious defects of the child's eyes, such facts were recorded. Cases of ophthalmia were not recorded as such, but some of its late effects were registered under the headings "Disease of Cornea, 68," and "Eye lost by disease, 70	836	637	764	692	2,929
4.	F	CHILDREN THE SUBJECTS OF RICKETS.—When bent legs or pigeon breast indicated Rickets the fact was recorded; conditions of the cranium were not accepted as evidence of Rickets, but were entered under their respective headings, see "Cranium (1 to 10) in Nomenclature of defects." Probably more children than those registered were or had been rachitic	157	39	39	9	244
5.	G	EXCEPTIONAL CHILDREN.—This includes all children whose physical or mental conditions show them to be obviously at a permanent disadvantage therefrom in social life. This group includes: Idiots (76); Imbeciles (77); "Children Feebly gifted Mentally (78); Children Mentally exceptional (79); Epileptics and children with history of fits during school life (80); Dumb children (81); and all children Crippled, Deformed, Maimed, Paralysed, see Group 11 and in Nomenclature (82). All these Exceptional Children need to be considered individually as to their special requirements	303	204	153	148	808
6.	G (76).	Idiots.—Includes all children who on account of their bodily and brain defects and the absence of mental power might be certified as idiots under the Idiots' Act and sent to an asylum. See Nomenclature (76) ...	2	...	...	...	2
7.	G (77).	Imbeciles.—This includes all children who might be certified as mentally imbecile and transferred to an asylum. Speaking generally, these are less hopeless cases than the idiots and more educible under industrial training. Some of these cases were the result of disease, not of congenital defect of brain. See Nomenclature (77)	30	16	3	2	51

* See note at foot of page 73.

Catalogue of Groups of Children and Groups of Cases—*Continued.*

Group.	Symbol.		Number of Children in which some defect was noted.				
			Of 50,000 seen in 1888—1891.		Of 50,000 seen in 1892—1894.		Of total 100,000 1888-94.
			Boys.	Girls.	Boys.	Girls.	B. & G.
8.	G (78).	"*Children Feebly-Gifted Mentally.*"—These children are distinctly deficient in mental power but might not be certified as imbeciles, and are therefore not fit for such medical certification: No child was registered in this group unless it was believed upon evidence observed and the teacher's report combined to be incapable of school work in the ordinary classes. It is difficult to define what physical conditions seen, as apart from mental tests, indicate the child as unfitted in mental capacity for the usual methods of education, and an arbitrary attempt to do so has not been made. There appears, however, to be a large class of "children feebly-gifted mentally," with defect of mental power short of imbecility but still with some deficiency. See Nomenclature.'(78)				
9.	G (79).	*Children Mentally Exceptional.*—These children while not necessarily mentally dull, and without brain power, appeared deficient in certain mental characteristics and in moral sense, such as, habitual liars, thieves and incendiaries; others were liable to attacks of total mental confusion, or periods of total mental inaptitude or violent passion, or were moral imbeciles. Some of these children were the offspring of insane parents or criminals. It is quite possible that some of these children were really epileptic or subject to *petit-mal.* Some of these children while thus mentally exceptional were not ordinarily dull pupils in schools. See Nomenclature (79)	89	85	49	52	275
10.	G (80).	*Epileptics and Children with History of Fits during School Life.*—In every school enquiry was made for children subject to fits, whether occurring in school or alleged to occur at home during school life and given as a reason for absence from school. A report given as to history of fits was recorded and the case was entered in this group, but at the	3	9	4	3	19

11. G (82).	inspection of a school, facts could not be usually observed proving the child to be epileptic. See Nomenclature (80)	36	18	21	35	110
	Children Crippled, Maimed, Deformed, or Paralysed.—Any child crippled, maimed, deformed, or paralysed was included in this group. Conditions of disease and paralysis were in various stages, but in all cases the child appeared to be at some permanent disadvantage. The conditions causing crippling were in various stages; many of these children were quite capable of work and play, some were mentally defective, they varied greatly in brain power and in physical health. Eye cases are not included in this group. See Nomenclature (82) ...					
12.	*Children who appear to require special Care and Training.*—This group includes all cases given as "Exceptional Children," see Group 5, and in addition "all children mentally dull, with defects in Development, abnormal Nerve-signs and Low Nutrition, *i.e.* Group 27	155	84	75	60	374
		473	344	226	218	1,261

CASES NOTED—ARRANGED ACTUARIALLY.

A large number of Groups of cases may be arranged actuarially for the purposes of scientific classification and research; they vary much in number and apparently in social importance also. There are Four Main Classes of Defectiveness:

A+ *Defects in Development* of the body and its parts in size, form or proportions of parts. See Nomenclature (1 to 42).

B+ *Abnormal Nerve-signs.*—Certain abnormal actions, movements, and balances which are described in the Nomenclature (43 to 63).

C+ *Low Nutrition* as indicated by the child being thin, pale, or delicate.

D+ *Mental Dulness.*—The Teacher's report as to mental ability was added to the record of every child registered, and those stated to be below the average in ability for school work were registered as "Dull."

For the purposes of actuarial research all cases registered have been arranged in 16 Primary groups. Cases in each Primary group present only the Main class of defects indicated A to D, but might also include E, F, or G, as explained on page 17. The cases with defects E, F, or G, would be the last as a Primary group, though not set out in the catalogue.

Groups 13 to 27 (seen 1888—91) have been calculated as closely as possible, but could not be accurately counted.

Catalogue of Groups of Children and Groups of Cases—*Continued.*

Group.	Symbol.	Primary Groups of Defects.	Of 50,000 seen in 1888—1891. Boys.	Girls.	Of 50,000 seen in 1892—1894. Boys.	Girls.	Of total 100,000 1888-94. B. & G.
13.	A	*Development cases only.*—Each case presents one or more defects in Development (1 to 42), but has no Abnormal Nerve-sign, no Low Nutrition, and is not Mentally dull	856	469	802	445	2,572
14.	B	*Nerve-cases only.*—Each case presents one or more Abnormal Nerve sign, see Nomenclature (43 to 63), but has no defect in Development, and is not of Low Nutrition or Mentally dull	751	441	1,059	762	3,013
15.	C	*Low Nutrition cases only.*—Each case is thin, pale, or delicate, but without defect in Development, Nerve-signs, or Mental dulness	34	10	108	110	262
16.	D	*Dull cases only.*—Each case is reported as Mentally dull without either defect in Development, Abnormal Nerve-sign, or Low Nutrition	243	110	331	297	981
17.	A B	*Development cases with Abnormal Nerve-signs only.*—Each case presents one or more defect in Development, and one or more Abnormal Nerve-sign, but neither Low Nutrition nor Mental dulness	920	397	415	207	1,939
18.	A C	*Development cases with Low Nutrition only.*—Each case presents one or more defect in Development with Low Nutrition, but neither Abnormal Nerve-sign nor Mental dulness	222	212	134	162	730

No.	Code					
19.	A D	*Development cases: Dull only.*—Each case presents one or more defect in Development with Mental dulness, but neither Abnormal Nerve-signs nor Low Nutrition	464	316	394	314	1,488
20.	B C	*Nerve cases with Low Nutrition only.*—Each case presents one or more Abnormal Nerve-sign and Low Nutrition, but neither defect in Development nor Mental dulness	152	132	115	109	508
21.	B D	*Nerve cases: Dull only.*—Each case presents one or more Abnormal Nerve-sign and is Mentally dull, but presents neither defect in Development nor Low Nutrition	444	320	703	487	1,954
22.	C D	*Low Nutrition cases: Dull only.*—Each case presents Low Nutrition and Mental dulness, but neither defect in Development nor Abnormal Nerve-sign	40	20	63	53	176
23.	A B C	*Development cases with Abnormal Nerve-signs and Low Nutrition only.*—Each case presents one or more defect in Development and one or more Abnormal Nerve-sign with Low Nutrition, but not Mentally dull.	220	224	69	77	590
24.	A B D	*Development cases with Abnormal Nerve-signs and Dull only.*—Each case presents one or more defect in Development and one or more Abnormal Nerve-sign with Mental dulness, but not Low Nutrition	643	318	323	224	1,508
25.	A C D	*Development cases with Low Nutrition and Dull only.*—Each case presents one or more defect in Development with Low Nutrition and Mental dulness without any Abnormal Nerve-sign	99	133	91	110	433
26.	B C D	*Nerve cases with Low Nutrition and Dull only.*—Each case presents one or more Abnormal Nerve-sign with Low Nutrition and Mental dulness without any defect in Development	71	85	89	70	315
27.	A B C D	*Development cases with Abnormal Nerve-signs, Low Nutrition, and Mental Dulness.*—Each case presents one or more defect in Development, and one or more Abnormal Nerve-sign with Low Nutrition and Mental dulness	192	157	80	79	508

The "Compound groups of cases" (28 to 56) are those presenting the class or classes of defects indicated, either alone or in combination with other defects. The term is used in contradistinction to Primary groups of cases (13 to 27) which present the defect or defects indicated not combined with others. The method of obtaining compound groups is by the addition of all the primary groups containing the defect or combination of defects indicated; these are set out in columns in the Catalogue.

Catalogue of Groups of Children and Groups of Cases—*Continued.*

The largest Groups of cases are those containing all Children presenting the same main Class of Defect either alone or in combination.

A+. *All Development cases.*—Each case presents one or more defect in Development, see Nomenclature (1) to (42), either with or without Nerve-signs, Low Nutrition, or Mental dulness. In each of these children some defect of body in growth, size, proportioning of parts, or development of tissue was observed ...

Group.	Defects Included.	Symbol.		Number of Children in which some defect was noted.				
				Of 50,000 seen in 1888—1891.		Of 50,000 seen in 1892—1894.		Of total 100,000 1888—94.
				Boys.	Girls.	Boys.	Girls.	B. & G.
28.	A AB AC AD ABC ABD ACD ABCD	A+.		3,616	2,235	2,308	1,618	9,777

No.		Description					Total
29.	B +.	*All Nerve cases.* Each case presents one or more Abnormal Nerve-signs, see Nomenclature (43) to (63), either with or without some defect in Development, Low Nutrition or Mental dulness. Some of the Nerve-signs indicate over-mobility, others, want of due action or response to stimulation, or want of due co-ordination from efficient training. Other signs indicate defectiveness in organisation of the brain … … …	3,413	2,074	2,853	2,015	10,355
		B A B B C B D A B C A B D B C D A B C D					
30.	C +.	*All Low Nutrition cases.*—Each case is thin, pale or delicate, either with or without some defect in Development, Nerve-sign, or Mental dulness … … … … … …	1,030	973	749	770	3,522
		C A C B C C D A B C A C D B C D A B C D					
31.	D +.	*All Dull cases.*—Each case is reported by the teachers as Dull Mentally or below the average in ability at school work, either with or without some defect in Development, Nerve-sign or Low Nutrition. All grades of Mental dulness were entered in this group. See Nomenclature (76) (77) (78) … … …	2,216	1,463	2,077	1,635	7,391
		D A D B D C D A B D A C D B C D A B C D					
		Groups of Cases with two Main Classes of Defects, either alone or in combination.					
32.	AB +.	*All Development cases with Abnormal Nerve-signs.*—Each case presents defect in development and one or more Nerve-signs, either alone or in combination with Low Nutrition, Mental dulness or both … … … … …	1,975	1,096	887	587	4,515
		A B A B C A B D A B C D					

Catalogue of Groups of Children and Groups of Cases—*Continued.*

Group.	Defects included.	Symbol.		Number of Children in which some defect was noted.				
				Of 50,000 seen in 1888—1891.		Of 50,000 seen in 1892—1894.		Of total 100,000 1888—94.
				Boys.	Girls.	Boys.	Girls.	B. & G.
33.	A C A B C A C D A B C D	AC+.	*All Development cases with Low Nutrition.*—Each case presents defect in Development with Low Nutrition, either alone or in combination with Nerve-signs, Mental dulness, or both	733	726	374	428	2,261
34.	A D A B D A C D A B C D	AD+.	*All Development cases with Mental Dulness.*—Each case presents defect in Development with Mental dulness, either alone or in combination with Nerve-signs, Mental dulness, or both	1,398	928	888	727	3,941
35.	B C A B C B C D A B C D	BC+.	*All Nerve cases with Low Nutrition.* Each case presents one or more Nerve-signs with Low Nutrition, either alone or in combination with defects in Development, Mental dulness, or both...	635	598	363	335	1,921
36.	B D A B D B C D A B C D	BD+.	*All Nerve cases with Mental Dulness.*—Each case presents Nerve-signs with Mental dulness, either alone or in combination with defect in Development, Low Nutrition, or both	1,370	880	1,195	860	4,305
37.	C D A C D B C D A B C D	CD+.	*All Low Nutrition cases with Mental Dulness.*—Each case presents Low Nutrition with Mental dulness, either alone or in combination with defect in Development, one or more Nerve-sign, or both ...	402	395	323	312	1,432

Groups of Cases with Three Main Classes of Defects, either alone or in combination.

No.	Class	Description					Total
38.	ABC / ABCD	**ABC+.** *All Development cases with Abnormal Nerve-signs and Low Nutrition.*—Each case presents one or more defect in Development with one or more Nerve-sign and Low Nutrition either alone or in combination with Mental dulness ...	412	381	149	156	1,098
39.	ABD / ABCD	**ABD+.** *All Development cases with Abnormal Nerve-signs and Mental dulness.*—Each case presents one or more defect in Development with one or more Nerve-sign and Mental dulness either alone or in combination with Low Nutrition ...	835	475	403	303	2,016
40.	ACD / ABCD	**ACD+.** *All Development cases with Low Nutrition and Mental Dulness.*—Each case presents one or more defect in Development with Low Nutrition and Mental dulness either alone or in combination with Nerve-sign ...	291	290	171	189	941
41.	BCD / ABCD	**BCD+.** *All Nerve-cases with Low Nutrition and Mental dulness.*—Each case presents one or more Nerve-sign with Low Nutrition and Mental dulness either alone or in combination with defect in Development ...	263	242	169	149	823

Groups of Cases with One Main Class of Defect without at least one other Main Class of Defect.

No.	Class	Description					Total
42.	A / C / AD / ACD	**(A+)—(AB+).** *All Development cases without Nerve-signs.*—Each case presents one or more defect in Development without any Nerve-sign, either alone or in combination with Low Nutrition, Mental dulness, or both ...	1,641	1,139	1,421	1,031	5,232
43.	A / AB / AD / ABD	**(A+)—(AC+).** *All Development cases without Low Nutrition.*—Each case presents one or more defect in Development without Low Nutrition, either alone or in combination with Nerve-sign, Mental dulness, or both ...	2,883	1,509	1,702	1,076	7,170

Catalogue of Groups of Children and Groups of Cases—*Continued.*

Group.	Defects included.	Symbol.	Of 50,000 seen in 1888—1891. Boys.	Girls.	Of 50,000 seen in 1892—1894. Boys.	Girls.	Of total 100,000 1888-94. B. & G.
44.	A AB AC ABC	(A +)—(A D +). *All Development cases without Mental Dulness.*—Each case presents one or more defect in Development without Mental dulness, either alone or in combination with Nerve sign, Low Nutrition, or both	2,218	1,307	1,420	891	5,836
45.	B BC BD BCD	(B +)—(A B +). *All Nerve-cases without Development defect.*—Each case presents one or more Nerve-sign without any defect in Development, either alone or in combination with Low Nutrition, Mental dulness, or both	1,438	978	1,966	1,428	5,810
46.	B AB BD ABD	(B +)—(B C +). *All Nerve-cases without Low Nutrition.*—Each case presents one or more Nerve-sign without Low Nutrition, either alone or in combination with defect in Development, Mental dulness, or both	2,778	1,476	2,500	1,680	8,434
47.	B AB BC ABC	(B +)—(B D +).—*All Nerve-cases without Mental Dulness.*—Each case presents one or more Nerve-sign without Mental dulness, either alone or in combination with defect in Development, Low Nutrition, or both	2,043	1,194	1,658	1,155	6,050
48.	C BC CD BCD	(C +)—(A C +). *All cases of Low Nutrition without defect in Development.*—Each case presents Low Nutrition without defect in Development, either alone or in combination with Nerve-signs, Mental dulness, or both	297	247	375	342	1,261
49.	C AC CD ACD	(C +)— (B C +). *All cases of Low Nutrition without Abnormal Nerve-signs.*—Each case presents Low Nutrition without Nerve-signs, either alone or in combination with defect in Development, Mental dulness, or both ...	395	375	396	435	1,601

No.	Code	Description					
50.	C / AC / BC / ABC	(C +)—(CD +). *All cases of Low Nutrition without Mental dulness.*—Each case presents Low Nutrition without Mental dulness, either alone or in combination with defect in Development, one or more Nerve sign, or both	628	578	426	458	2,000
51.	D / BD / CD / BCD	(D +)—(AD +). *All cases of Mental dulness without defect in Development.*—Each case presents Mental dulness without defect in Development, either alone or in combination with Nerve-sign, Low Nutrition, or both	818	535	1,186	907	3,446
52.	D / AD / CD / ACD	(D +)—(BD +). *All cases or Mental dulness without Abnormal Nerve-signs.*—Each case presents Mental dulness without Nerve-signs, either alone or in combination with defect in Development, Low Nutrition, or both	846	583	879	774	3,082
53.	D / AD / BD / ABD	(D +)—(CD +). *All cases of Mental dulness without Low Nutrition.*—Each case presents Mental dulness without Low Nutrition, either alone or in combination with defect in Development, Nerve-sign, or both	1,814	1,068	1,751	1,322	5,955
54.	C / D / CD	*All cases without defect in Development and without Nerve-signs.*—Each case presents no defect in Development and no Nerve-sign, but may present Low Nutrition, Mental dulness, or both	337	140	502	460	1,439
55.	D / C D	*All cases without defect in Development and without Nerve-sign, but Mentally dull.*—Each case presents Mental dulness but no defect in Development and no Nerve-sign, either with or without Low Nutrition	303	164	394	350	1,211
56.	E, F or G	*All cases without either defect in Development, Nerve-sign, Low Nutrition, or Mental dulness.*—These cases present none of the four main classes of defects; they belong to the classes E, Eye cases; F, Rickets; or G, Exceptional Children (79) to (103)	208	254	336	323	1,121

TABLE IX. (*cases seen* 1888-91).—*Distribution of Signs or Defects under Divisions of Schools.*

For Definition of Signs, see Nomenclature on page 72.

For Description of Divisions of Schools, see List of Schools on page 5.

For Further Distribution in Divisions of Schools, see sub-table XX.

Numbers refer to Nomenclature.	1. Poor Law Schools.		2. Certified Industrial.		3. Homes and Orphanages.		4. Public Elementary Day Schools.	
	Boys.	Girls.	Boys.	Girls.	Boys.	Girls.	Boys.	Girls.
A. DEFECTS IN DEVELOPMENT.								
(a 1). CRANIUM defective.........	387	171	160	40	34	79	947	758
(a 2). CRANIUM large	76	19	7	1	4	2	170	24
(a 3). CRANIUM small	42	67	24	26	14	53	247	592
(a 4). CRANIUM bossed...........	131	24	49	3	12	16	305	82
(a 5). Forehead defective.........	80	41	43	7	2	3	58	27
(a 6). Frontal ridge	7	4	4	1	...	1	78	21
(a 7). CRANIUM asymmetrical...	22	6	9	1	1	3	52	6
(a 8). Dolichocephalic	7	8	7	1	...	...	28	2
(a 10). Other types of CRANIUM..	22	2	17	...	1	1	10	3
(b 11). EXTERNAL EAR defective	257	81	106	3	41	9	643	175
(c 12). EYELIDS WITH EPICAN-THIS	124	101	28	3	14	18	348	262
(d 13). PALATE defective in shape	216	133	67	18	32	30	448	344
(d 14). PALATE narrow	114	73	35	3	19	17	292	198
(d 15). PALATE V-shaped	85	54	6	4	10	12	134	101
(d 16). PALATE arched	15	5	22	11	2	...	47	25
(d 17). PALATE cleft	2	1	1	...	1	1	11	5
(d 18). Other defects of PALATE.	...	...	3	...	...	...	7	15
(e 19). NASAL BONES defective...	...	...	...	...	...	...	...	...
(f 20). GROWTH SMALL, short ...	41	48	21	6	3	9	144	146
(g 21). OTHER DEFECTS IN DEVELOPMENT	254	160	122	21	27	38	505	426
(g 22). Adipose type	3	9	7	...	...	...	6	...
(g 26). Face small................	2	3	3	...	...	...	13	8
(g 27). Features coarse	69	42	28	10	3	6	47	46
(g 28). Forehead hairy	20	3	11	...	...	...	9	...
(g 29). Frontal Veins large........	4	3	...	...	...	...	9	3
(g 30). Hands blue and cold	13	7	1	1	...	...	3	3
(g 31). Hare lip.................	2	1	...	...	...	...	8	4
(g 32). Icthyosis	7	3	...	...	...	...	3	4
(g 34). Mouth small..............	1	2	3	1	1	...	22	14
(g 36). Nose soft tissue	3	2	2	...	...	...	28	26
(g 39). Palpebral fissures	14	17	16	5	5	4	63	57
(g 40). Prognathous type	3	...	7	...	1	...	11	...
B. ABNORMAL NERVE SIGNS.								
(43). General balance	74	32	12	6	...	1	115	134
(44). Expression defective	259	146	56	21	17	31	362	276
(45). Frontal overacting	423	107	175	14	43	28	681	145
(46). Corrugation	38	4	28	5	4	6	129	25
(47). Orbicularis oculi relaxed..	121	66	30	5	12	10	359	262
(48). Eye Movements defective.	120	75	87	11	37	32	554	367

	1. Poor Law Schools.		2. Certified Industrial.		3. Homes and Orphanages.		4. Public Elementary Day Schools.	
	Boys.	Girls.	Boys.	Girls.	Boys.	Girls.	Boys.	Girls.
ABNORMAL NERVE SIGNS, *con.*								
(49). Head Balance	66	47	13	8	1	2	139	262
(50). Hand Balance weak	189	66	49	14	47	41	430	383
(51). Hand Balance nervous ...	93	55	35	5	2	9	420	447
(52). Finger twitches	80	25	25	3	2	8	338	225
(53). Lordosis.......................	36	45	14	4	2	6	132	224
(*h* 54). OTHER ABNORMAL NERVE SIGNS	144	71	34	13	8	13	248	137
(55). Deaf, or Hearing defective	15	9	3	1	...	4	16	19
(56). Grinning	29	15	8	2	...	4	32	22
(57). Mouth open	31	19	15	2	6	1	82	37
(58). Over-Mobile	1	...	...	...	...	...	3	...
(59). Response in action defective	25	11	4	4	2	4	81	37
(60). Speech defective	41	22	4	1	...	6	71	41
(61). Statuesque or Immobile...	...	3	...	2	...	...	2	...
(62). Tremor	14	7	3	...	...	...	13	4
E. DEFECTS OF EYES.								
(64). Squint	128	69	25	12	15	19	317	222
(65). Using convex glasses	16	10	...	...	...	2	108	140
(66). Using concave glasses......	5	2	1	...	2	...	17	21
(67). Myopia, not using glasses.	3	2	...	1	...	...	5	8
(68). Disease of cornea............	17	9	12	4	4	2	42	39
(69). Eye lost by accident	5	8	1	...	...	...	12	10
(70). Eye lost by disease.........	9	4	1	...	6	4	2	4
(71). Nystagmus	6	3	...	...	...	...	19	6
(72). Ptosis	10	2	6	...	1	...	9	9
(73). Pupils unequal...............	1	3	...	...	...	...	1	...
(75). Miscellaneous defects......	2	...	8	3	...	...	10	13
F. CASES OF RICKETS	38	12	5	...	1	4	113	23
G. EXCEPTIONAL CHILDREN.								
(82). Crippled, maimed, &c. ...	64	21	5	2	19	18	67	43
(83). Disease of hip	9	3	...	1	2	3	13	11
(84). Disease of spine	10	5	2	1	4	5	9	6
(85). Disease of upper limb......	7	1	1	...	1	...	2	1
(86). Disease of lower limb......	4	1	1	...	1	2	7	4
(87). Hand maimed	...	...	...	...	...	...	...	...
(88). Amputation of arm.........	...	...	...	...	...	...	...	1
(89). Amputation of leg	7	...	...	...	4	3	4	5
(90). Congenital absence of greater part of upper limb	1	1	...	...	...	...	2	1
(91). Congenital absence of hand	...	1	...	...	...	1	...	2
(92). Congenital defect of hand.	2	...	...	...	...	...	2	2
(93). Congenital absence of foto	...	...	...	...	...	...	...	1
(94). Club foot	...	...	...	...	...	...	...	...
(95). Hemiplegia	7	2	...	...	...	2	7	5
(96). Paraplegia	...	1	...	...	2	1	...	...
(97). Infantile palsy, upper limb	3	2	...	...	...	...	7	2
(98). Infantile palsy, lower limb	13	4	1	...	5	...	13	2
(99). Torticollis	...	...	...	...	...	1	1	...

96

TABLE X. (*cases seen* 1888-91).—*Distribution of Groups of Children under Divisions of Schools.*

For Definition of Groups of Children, see Catalogue on page 82.
For Description of Divisions of Schools, see List of Schools on page 5.
For Further Distribution of Divisions of Schools, see sub-table XX.

Numbers refer to Catalogue.	1. Poor Law Cases.		2. Certified Industrial.		3. Homes and Orphanages.		4. Public Elementary Day Schools.	
	Boys.	Girls.	Boys.	Girls.	Boys.	Girls.	Boys.	Girls.
NUMBER OF CHILDREN SEEN	5857*	3,947	1,588	407	774	1,049	18,638	17,740
1. Normal children	4,552	3,262	1,088	316	602	863	15,063	15,095
2. Children noted and registered	1,332	685	500	91	172	186	3,575	2,645
3. Eye cases	205	112	54	20	28	27	549	478
4. Children with Rickets	38	12	5	...	1	4	113	23
5. Exceptional children	109	61	5	8	21	32	168	103
6. Idiots	...	...	...	...	...	...	2	...
7. Imbeciles	12	7	...	...	...	2	18	7
8. Children feebly gifted mentally	32	33	...	4	2	11	55	37
9. Children mentally exceptionl	...	1	...	2	...	...	3	6
10. Epileptics	4	1	...	...	...	1	32	16
11. Crippled, maimed, paralysed.	64	21	5	2	19	18	67	43
12. Children who appear to require special training	137	74	16	14	24	41	296	215
PRIMARY GROUPS OF CASES.								
13. Development cases only	189	127	65	9	2	17	600	316
14. Nerve cases only	172	70	66	...	17	18	496	353
15. Low Nutrition cases only	15	7	...	...	...	...	19	3
16. Dull cases only	50	15	13	3	...	17	180	75
17. Developm'nt cases with Nerve signs only	262	108	101	11	47	24	510	254
18. Development cases with Low Nutrition only	46	20	6	5	6	8	164	179
19. Development cases who were Dull only	92	67	57	7	19	25	296	217
20. Nerve cases with Low Nutrition only	47	6	5	5	1	3	99	118
21. Nerve cases Dull only	95	54	52	12	16	18	281	236
22. Low Nutrition cases Dull only	3	10	1	...	2	...	34	10
23. Developm'nt cases with Nerve signs & Low Nutrition only	61	28	9	8	2	4	148	184
24. Developm'nt cases with Nerve signs and Dull only	186	93	79	12	26	29	352	184
25. Development cases with Low Nutrition and Dull only	18	13	1	4	3	7	77	109
26. Nerve cases with Low Nutrition and Dull only	12	10	9	3	1	5	49	67
27. Developm'nt cases with Nerve signs, Low Nutrition, and Mental Dulness	34	19	11	6	2	11	145	121
THE LARGEST GROUPS OF CASES IN THE MAIN CLASSES OF DEFECTS.								
28. All Development cases	888	475	329	62	107	134	2,292	1,564
29. All Nerve cases	889	388	332	57	112	112	2,080	1,517
30. All Low Nutrition cases	236	93	41	28	14	35	739	817
31. All Mentally Dull cases	510	281	223	47	68	113	1,415	1,022

* See note on page 73.

	1. Poor Law Schools.		2. Certified Industrial.		3. Homes and Orphanages.		4. Public Elementary Day Schools.	
THE LARGEST GROUPS OF CASES IN TWO MAIN CLASSES OF DEFECTS.	Boys.	Girls.	Boys.	Girls.	Boys.	Girls.	Boys.	Girls.
32. All Development cases with Abnormal Nerve signs	543	248	200	37	77	68	1,155	743
33. All Development cases with Low Nutrition	159	80	27	23	13	30	534	593
34. All Development cases with Mental Dulness	330	192	148	29	50	76	870	631
35. All Nerve cases with Low Nutrition	154	63	34	22	6	23	441	490
36. All Nerve cases with Mental Dulness	347	176	151	33	45	63	827	608
37. All Low Nutrition cases with Mental Dulness	67	52	22	13	8	20	305	310
THE LARGEST GROUPS OF CASES IN THREE MAIN CLASSES OF DEFECTS.								
38. All Development cases with Nerve signs & Low Nutrition	95	47	20	14	4	15	293	305
39. All Development cases with Nerve signs and Mental Dulness	220	112	90	18	28	40	497	305
40. All Development cases with Low Nutrition and Dulness	52	32	12	10	5	18	222	230
41. All Nerve cases with Low Nutrition & Mental Dulness	46	29	20	9	3	16	194	188
THE LARGEST GROUPS OF CASES IN ONE MAIN CLASS OF DEFECTS, AND ABSENT FROM AT LEAST ONE OTHER MAIN CLASS OF DEFECTS.								
42. All Development cases without Nerve signs	345	227	129	25	30	66	1,137	821
43. All Development cases without Low Nutrition	729	395	302	39	94	104	1,758	971
44. All Development cases without Mental Dulness	558	283	181	33	57	58	1,422	933
45. All Nerve cases without defect in Development	346	140	132	20	35	44	925	774
46. All Nerve cases without Low Nutrition	735	325	298	35	106	89	1,639	1,027
47. All Nerve cases without Mental Dulness	542	212	181	24	67	49	1,253	909
48. All Low Nutrition cases without Defect in Development	77	13	14	5	1	5	205	224
49. All Low Nutrition cases without Nerve signs	82	30	7	6	8	12	298	327
50. All Low Nutrition cases without Mental Dulness	169	41	19	15	6	15	434	507
51. All Mentally Dull cases without Development Defect	180	89	75	18	18	37	545	391
52. All Mentally Dull cases without Nerve signs	163	105	72	14	23	50	588	414
53. All Mentally Dull cases without Low Nutrition	443	229	201	34	60	93	1,110	712
54. All cases without Development or Nerve Defect								
55. All cases without Developm't or Nerve Defect but Dull	44	22	13	5	7	7	239	130

G

TABLE XI. *(Cases seen 1888-91).*—*Conditions of Defective Development in Relation to Low Nutrition, Mental Dulness, and Nerve Defects.*

Among 2,794 Boys, 2,550 Girls. January, 1889.			Low Nutrition.		Mental Dulness.		Nerve Defects.	
			B.	G.	B.	G.	B.	G.
Total of cases presenting some defects of development, including cranial abnormalities, palate, ears epicanthis, and other defects (not including squint) boys, 274; girls, 125; total, 399	62	40	86	39	101	44
Cranial defects alone, not in combination	22	12	33	11	31	12
Palate defective alone	6	3	5	12	9	3
Ears defective alone	5	4	8	3	11	4
Epicanthis folds alone	1	5	3	1	1	2
The 73 cases with defects other than those mentioned presented	10	4	18	4	23	4
Cases of Binary Defects.						
Defects of cranium and palate	12	4	12	4	20	7
,, ,, ,, ears	6	0	11	0	12	0
,, ,, ,, epicanthic folds	—		1	0	3	2
,, ,, ,, other defects than those mentioned	5	3	6	2	6	2
,, palate and ears	2	0	7	0	6	0
,, ,, ,, epicanthic folds	1	0	2	2	4	2
,, ,, ,, other defects	0	1	4	2	5	2
,, ears and epicanthis	1	1	3	2	3	0
,, ,, other defects	2	0	5	0	1	0

	B.	G.	Low Nutrition.		Mental Dulness.		Nerve Defects.	
Cases of Triple Defects.	24	7	11	6	14	2	11	6
The Palate was examined in 459 cases:								
It was found normal in ...	265	77	—		—		—	
,, ,, abnormal in ...	77	40	—		—		—	
Defects of palate:								
Arched, narrow, high, or vaulted	68	37						
V-shaped, not included above ...	6	2	29	13	22	20	29	16
Of the flat type ...	3	1						
	77	40						
Defects of ears:								
Symmetrical	37	13	12	2	13	2	15	3
Asymmetrical	27	4	5	2	11	2	9	2
	64	17						
Epicanthic folds:								
Symmetrical or double ...	27	17	5	4	9	6	6	3
Single or most marked on one side	10	4	1	1	1	0	1	0
	37	21						

TABLE XII. (*Cases seen 1888-91*).—*Binary Combinations of Defects in Development of the body.*

Total of Cases presenting each Sign respectively, among 2,754 Boys, 2,650 Girls.—Jan., 1889.				Cranial Abnormalities.			Palate Defects.			Ear Defects.			Epicanthic fold.			Other Defects.			No other Defect.		
	B.	G.	T.	B.	G.	T.	B.	G.	T.	B.	G.	T.	B.	G.	T.	B.	G.	T.	B.	G.	T.
Cranial Abnormalities.																					
In 10 Public Elementary Schools ...	82	37	119		X		18	8	26	11	1	12	2	2	4	2	5	7	62	25	87
„ 4 Special Schools	84	28	112				24	5	29	12	0	12	2	0	2	10	1	11	20	17	46
	166	65	231																		
Palate Defective.																					
„ 10 Public Elementary Schools ...	35	18	53	18	8	26		X		6	0	6	4	5	9	2	0	2	11	7	18
„ 4 Special Schools	42	22	64	24	5	29				10	0	10	3	0	3	8	5	13	11	13	24
	77	40	117																		
Ears Defective.																					
„ 10 Public Elementary Schools ...	36	13	49	11	1	12	6	0	6		X		4	3	7	4	0	4	19	8	27
„ 4 Special Schools	28	4	32	12	0	12	10	0	10				2	0	2	5	0	5	9	4	13
	64	17	81																		
Epicanthic fold																					
„ 10 Public Elementary Schools ...	24	19	43	2	2	4	4	5	9	4	3	7		X		5	0	5	13	11	24
„ 4 Special Schools	13	2	15	2	0	2	3	0	3	2	0	2				3	0	3	6	2	8
	37	21	58																		
Other Defects than those above.																					
„ 10 Public Elementary Schools ...	37	10	47	2	5	7	2	0	2	4	0	4	5	0	5		X		11	5	16
„ 4 Special Schools	20	6	26	10	1	11	8	5	13	5	0	5	3	0	3				12	1	13
	57	16	73																		

TABLE XIII. (*Cases seen 1888-91*).—*Showing the number of children with each Nerve-sign; the numbers with each Combination of these Signs; the number of children in whom each Sign occurred alone; and the number of times each Sign respectively occurred in Combination.*

Each cell is given as Boys, Girls, Total.

	Total of Cases presenting each Nerve-Sign Respectively, among 2,794 Boys, 2,550 Girls. (1889).	Nervous Hand	Weak Hand	Lordosis	Frontals Over-acting	Orbicular Muscle of Eyelids Toneless	Finger Twitches	Nerve-signs not in combination
Nervous Hand		×						
In 10 Public Elementary Schools	44, 69, 113	×	—, —, —	10, 28, 38	9, 1, 10	5, 4, 9	15, 16, 31	21, 28, 49
4 Special Schools	25, 5, 30	×	—	6, 2, 8	7, 0, 7	—	6, 2, 8	11, 3, 14
(subtotal 69, 74, 143)								
Weak Hand			×					
10 Public Elementary Schools	10, 22, 32	—, —, —	×	1, 6, 7	3, 0, 3	1, 3, 4	3, 5, 8	5, 9, 14
4 Special Schools	16, 6, 22	—	×	3, 3, 6	6, 1, 7	2, 1, 3	3, 3, 6	5, 2, 7
(subtotal 26, 28, 54)								
Lordosis				×				
10 Public Elementary Schools	15, 41, 56	10, 28, 38	1, 6, 7	×	3, 1, 4	0, 2, 2	1, 10, 11	4, 6, 10
4 Special Schools	18, 5, 23	6, 2, 8	3, 3, 6	×	3, 0, 3	1, 0, 1	6, 1, 7	6, 2, 8
(subtotal 33, 46, 79)								
Frontals Overacting.					×			
10 Public Elementary Schools	41, 2, 43	9, 1, 10	3, 0, 3	3, 1, 4	×	2, 0, 2	3, 0, 3	25, 0, 25
4 Special Schools	68, 13, 81	7, 0, 7	6, 1, 7	3, 0, 3	×	—	8, 2, 10	48, 10, 58
(subtotal 109, 15, 124)								
Orbicular Muscle of Eyelids Toneless.						×		
10 Public Elementary Schools	14, 11, 25	5, 4, 9	1, 3, 4	0, 2, 2	2, 0, 2	×	3, 3, 6	9, 3, 12
4 Special Schools	4, 1, 5	—	2, 1, 3	1, 0, 1	—	×	—	1, 0, 1
(subtotal 18, 12, 30)								
Finger Twitches.							×	
10 Public Elementary Schools	20, 27, 47	15, 16, 31	3, 5, 8	1, 10, 11	3, 0, 3	3, 3, 6	×	3, 3, 6
4 Special Schools	25, 10, 35	6, 2, 8	3, 3, 6	6, 1, 7	8, 2, 10	—	×	8, 5, 13
(subtotal 45, 37, 82)								
Total, 10 Public Elementary Schools	…	39, 49, 88	8, 14, 22	15, 47, 62	20, 2, 22	11, 12, 23	25, 34, 59	67, 49, 116
4 Special Schools	…	19, 4, 23	14, 8, 22	19, 6, 25	24, 3, 27	3, 1, 4	23, 8, 31	79, 22, 101
Final Total	…	58, 53, 111	22, 22, 44	34, 53, 87	44, 5, 49	14, 13, 27	43, 42, 90	146, 71, 217

TABLE XIV. (*Cases seen 1888-91*).—*Cases that appear to require Special Training and Care on grounds of Physical or Mental Conditions.*

No. of Cases, Schools I.—CVI.			Classes of Cases composing this Group.	Poor Law Schools, I.—XIX.			Certified Industrial Schools, XX.—XXVIII.			Homes and Orphanages, XXIX.—XXXIV.			Public Elementary, XXXV.—CVI.		
Boys	Girls	Total		Boys	Girls	Total	Boys	Girls	Total	Boys	Girls	Total	Boys	Girls	Total
124	110	234	Cases exceptional in mental status. Groups 6, 7, 8, 9	44	41	85	...	6	6	2	13	15	78	50	128
36	18	54	Epileptics. Group 10	4	1	5	...	...	...	...	1	1	32	16	48
155	84	239	Crippled, paralysed, &c. Group 11	64	21	85	5	2	7	19	18	37	67	43	110
192	157	349	Cases defective in development with abnormal nerve signs and low nutrition, also reported as dull by teachers. Group 27	34	19	53	11	6	17	2	11	13	145	121	266
507	369	876	Total number of cases	146	82	228	16	14	30	23	43	66	322	230	552
472	344	816	Children who appear to require special care, i.e., mental status defective, epileptic, crippled, also children with "defects in development, abnormal nerve signs, and low nutrition, also reported as mentally dull by the teachers."	137	74	211	16	14	30	23	41	64	206	215	511

In the table above some children appear in more than one class; such overlapping cases have been allowed for in the second line of totals above, which give the actual number of children. These cases are arranged below.

No. of Cases, Schools I.—CVI.			Classes of Cases composing this Group.	Poor Law Schools, I.—XIX.			Certified Industrial Schools, XX.—XXVIII.			Homes and Orphanages, XXIX.—XXXIV.			Public Elementary, XXXV.—CVI.		
Boys	Girls	Total		Boys	Girls	Total	Boys	Girls	Total	Boys	Girls	Total	Boys	Girls	Total
4	1	5	"Cases defective in development with abnormal nerve signs and low nutrition, also reported dull by teachers," and crippled	2	...	2	...	...	...	...	...	...	2	1	3
17	13	30	Cases defective in development with abnormal nerve signs and low nutrition and mentally exceptional	4	5	9	...	...	...	...	2	2	13	6	19
1	2	3	"Cases defective in development with abnormal nerve signs and low nutrition, also reported dull by teachers," and epileptics	...	1	1	...	...	...	...	...	...	1	1	2
1	1	2	Group as last, but also mentally defective	...	...	...	...	...	...	...	...	...	1	1	2
5	3	8	Cases defective in mental status and paralysed or crippled	3	1	4	...	...	...	...	...	...	2	2	4
5	5	10	Defective in mental status and epileptic	...	1	1	...	...	...	...	...	...	5	4	9
2	...	2	Epileptic and crippled or paralysed, not dull	...	...	...	...	...	...	...	...	...	2	...	2

TABLE XV. (Cases seen 1888-91).—Defects as observed in co-relation with Development defects, Abnormal Nerve-signs, Low Nutrition, and Mental dulness, in Divisions of Schools respectively.

Development defects

Children seen, 1888-91.	Poor Law Schools. I. to XIX.						Certified Industrial Schools. XX.—XXVIII.						Homes and Orphanages. XXIX.—XXXIV.						Day Schools. XXXV.—CVI.					
	Nerve-signs.		Low Nutrition.		Dull.		Nerve-signs.		Low Nutrition.		Dull.		Nerve-signs.		Low Nutrition.		Dull.		Nerve-signs.		Low Nutrition.		Dull.	
	Boys	Girls	Boys	Girls	Boys	Girls	Boys	Girls	Boys	Girls	Boys	Girls	Boys	Girls	Boys	Girls	Boys	Girls	Boys	Girls	Boys	Girls	Boys	Girls
Total of Development cases	543	248	159	80	330	192	200	37	27	23	149	29	77	68	13	30	50	76	1,155	743	534	593	870	631
Cranial abnormalities	237	94	77	35	155	81	109	31	16	21	79	19	23	47	5	21	19	43	481	359	204	493	380	334
External Ear defective	165	43	49	18	90	80	62	2	11	0	46	2	11	3	2	1	16	6	317	80	135	50	188	65
Eyelids with Epicanthis	59	39	16	12	43	38	16	3	0	2	11	2	11	10	0	4	4	13	141	108	49	57	134	83
Palate defective	139	66	35	14	93	57	47	11	4	6	27	13	19	15	2	6	18	13	236	170	132	129	186	149
Other defects in Developm'nt	178	94	39	20	128	69	86	16	17	4	58	12	18	23	3	4	13	21	273	222	130	144	184	161

Abnormal Nerve-signs

Children seen, 1888-91.	Poor Law Schools. I. to XIX.						Certified Industrial Schools. XX.—XXVIII.						Homes and Orphanages. XXIX.—XXXIV.						Day Schools. XXXV.—CVI.					
	Defects in Development		Low Nutrition.		Dull.		Defects in Development		Low Nutrition.		Dull.		Defects in Development		Low Nutrition.		Dull.		Defects in Development		Low Nutrition.		Dull.	
	Boys	Girls	Boys	Girls	Boys	Girls	Boys	Girls	Boys	Girls	Boys	Girls	Boys	Girls	Boys	Girls	Boys	Girls	Boys	Girls	Boys	Girls	Boys	Girls
Total of Nerve cases	543	248	154	63	347	176	200	37	34	22	151	33	77	68	6	23	45	63	1,155	743	441	490	827	608
General Balance defective	62	23	13	6	43	14	8	3	3	1	8	4	0	3	0	0	0	3	68	57	31	49	48	51
Expression defective	196	111	55	19	132	79	51	14	8	5	28	13	9	26	7	7	9	18	237	178	127	115	200	143
Frontals overacting	190	55	60	16	158	40	87	11	14	6	84	10	14	16	1	4	18	16	292	63	137	38	286	70
Corrugation	24	2	8	0	15	3	14	1	3	0	17	3	2	1	0	1	1	3	58	12	34	5	54	6
Orbicularis oculi lax	85	52	27	8	52	35	26	6	0	1	7	3	9	7	0	1	5	9	241	162	85	102	144	111
Eye movements defective	79	52	18	13	40	35	66	3	7	1	40	8	14	22	0	7	8	17	341	218	105	129	241	162
Head Balance	49	31	18	10	25	19	3	7	0	5	2	4	2	1	0	0	0	2	69	118	28	91	55	104
Hand Balance weak	121	41	37	15	80	25	39	10	7	1	28	1	27	21	4	8	21	26	188	124	67	83	157	123
Hand Balance nervous	39	27	17	12	26	9	22	4	5	1	15	1	0	5	0	2	0	5	192	169	89	143	148	145
Finger twitches	40	10	14	10	18	6	13	3	1	2	13	1	1	5	2	0	1	4	148	82	75	81	111	67
Lordosis	19	27	8	11	13	21	11	6	4	1	8	1	1	1	0	0	0	1	61	76	24	75	51	62
Other abnormal Nerve-signs	90	50	25	10	69	42	26	4	6	4	18	10	5	8	0	4	5	3	161	78	73	44	141	72

TABLE XVI. (*Cases seen* 1888-91).—*Co-relation of defects in Development with Nerve-signs:
Low Nutrition and Mental Dulness.*

For Definition, and number of cases of Defects, see Nomenclature on page 72.   For Groups of Cases see Catalogue on page 82.

Percentages are taken upon the number of children presenting the defects indicated in the first column.

Defects in Development. See Nomenclature, page 72.	With Group 29, Abnormal Nerve-signs, alone or in combination.				With Group 30, Low Nutrition, alone or in combination.				With Group 31, Mental Dulness, alone or in combination.			
	Number of cases.		Per cent. of cases.		Number of cases.		Per cent. of cases.		Number of cases.		Per cent. of cases.	
	Boys.	Girls.	Boys.	Girls.	Boys.	Girls.	Boys.	Girls.	Boys.	Girls.	Boys.	Girls.
All Development cases, Group 28	1,975	1,096	54·6	49·0	733	726	20·2	32·5	1,398	928	38·3	41·5
(1) Cranial defects	850	531	55·6	50·6	392	480	25·7	45·8	634	477	41·4	45·5
(3) Cranium small......	177	372	54·1	50·4	151	399	46·1	54·1	165	353	50·4	47·8
(7) Cranium asymmetrical	40	7	47·6	43·7	18	3	21·4	1·87	35	6	41·6	37·5
(11) Ears defective	566	128	54·0	47·7	196	72	18·7	26·8	340	103	32·4	38·4
(12) Epicanthis	227	160	44·1	41·6	65	73	12·6	19·0	192	136	37·3	35·4
(13) Palate defective	441	262	55·4	49·9	173	155	21·7	29·5	324	232	40·7	44·2
(19) Nasal bones	131	95	54·3	44·3	16	19	6·6	8·8	87	77	36·1	36·0
(20) Growth small	119	110	56·9	52·6	88	101	42·1	48·3	78	79	37·3	37·8
(21) Other defects in development	555	355	61·1	55·0	189	172	20·8	26·5	383	263	42·0	40·5
(27) Features coarse	112	68	76·1	65·3	19	17	12·9	16·3	73	43	49·6	41·3
(34) Mouth small	16	10	59·2	58·8	8	2	29·6	11·7	8	10	29·6	58·8
(39) Palpebral fissures	61	57	62·2	68·6	22	16	22·4	19·2	41	39	41·8	47·0

TABLE XVII. (*Cases seen 1888-91*).—*Co-relation of Abnormal Nerve-signs with Defect in Development, Low Nutrition, and Mental Dulness.*

For definition and number of cases with Nerve-signs, see Nomenclature on page 72; and for Groups of cases, see Catalogue on page 82.

Percentages are taken upon the number of children presenting the Nerve-signs indicated in the first column.

Abnormal Nerve-signs. See Nomenclature on page 72.	With Group 28, Defect in Development, alone or in combination.				With Group 29, Low Nutrition, alone or in combination.				With Group 30, Mental dulness, alone or in combination.			
	Number of cases.		Per cent. of cases.		Number of cases.		Per cent. of cases.		Number of cases.		Per cent. of cases.	
	Boys.	Girls.	Boys.	Girls.	Boys.	Girls.	Boys.	Girls.	Boys.	Girls.	Boys.	Girls.
(43) All Nerve cases, Group 20	1,975	1,096	57·8	52·8	635	598	18·6	28·8	1,370	880	40·1	42·4
(44) General Balance defective	138	86	68·6	49·7	47	56	23·3	32·3	99	72	49·2	41·5
(45) Expression defective	493	329	71·0	69·4	191	146	27·5	30·8	369	253	53·1	53·3
(46) Frontals overacting	583	145	41·1	49·6	221	64	16·7	21·7	548	136	41·4	46·2
(46) Corrugation	105	22	52·7	55·0	45	6	22·6	15·0	91	21	45·7	52·5
(47) Orbicularis Oculi relaxed	361	224	69·1	65·3	112	112	21·4	32·6	208	158	39·8	46·0
(48) Eye Movements defective	500	298	62·6	61·4	130	150	16·2	30·9	329	222	41·1	45·7
(49) Head Balance drooped	151	178	68·9	55·8	46	109	21·0	34·1	97	145	44·2	45·4
(50) Hand Balance weak	375	196	52·4	38·8	115	107	16·0	21·2	286	178	40·0	35·3
(51) Hand Balance nervous	253	205	46·0	39·7	111	158	20·1	30·6	189	170	34·3	32·9
(52) Finger twitches	202	99	45·3	37·9	90	95	20·2	36·4	143	78	32·1	29·8
(53) Lordosis	92	107	50·0	38·3	36	87	19·5	31·1	72	85	38·9	30·5
(54) "Other Nerve-signs"	278	135	64·0	57·6	104	58	23·9	24·7	228	125	52·5	53·4
(56) Grinning	52	27	75·3	62·8	13	11	18·8	25·5	39	26	56·5	60·4
(57) Mouth open	107	38	79·8	64·4	41	20	30·5	34·0	68	31	50·6	52·5
(59) Response defective	69	38	61·6	67·8	29	17	25·8	30·3	75	39	66·9	69·6
(60) Speech defective	30	44	25·9	62·8	10	13	8·6	18·5	27	43	23·2	61·4
(71) Nystagmus	7	6	28·0	66·6	6	3	21·0	33·3	10	4	40·0	44·4

TABLE XVIII. (*cases seen 1888-91*).—*Co-relation of Binary defects in Development with Nerve-signs, Low Nutrition, and Mental Dulness, in Resident Schools and in Day Schools.*

For Definition of Defects, see Nomenclature on page 72.   For Description of Groups of Cases, see Catalogue on page 82.
Percentages are taken upon the number of cases presenting the combination of defects indicated in the first column.

Binary defects in Development. See Nomenclature, page 72.	Number of Cases				With Abnormal Nerve-signs, alone or in combination, Group 29						With Low Nutrition, alone or in combination, Group 30						With Mental Dulness, alone or in combination, Group 31					
	Resident B.	Resident G.	Day B.	Day G.	Resident B.	Resident G.	Day B.	Day G.	%B.	%G.	Resident B.	Resident G.	Day B.	Day G.	%B.	%G.	Resident B.	Resident G.	Day B.	Day G.	%B.	%G.
(1) Cranium & (13) Palate	116	45	111	79	77	23	58	41	59·4	51·6	17	13	47	44	28·1	46·0	64	21	50	35	50·2	45·1
,, & (21) Other defect	138	139	143	80	107	44	84	77	68·0	55·2	30	15	61	80	33·4	43·3	80	34	60	61	49·8	43·3
,, & (11) Ear	91	19	103	25	61	10	55	8	60·0	40·9	14	8	34	14	24·7	50·0	33	9	35	10	35·0	43·4
,, & (12) Epicanthis	30	22	40	40	17	10	24	25	58·5	56·4	6	6	12	22	25·7	45·1	14	9	23	13	52·8	35·4
(13) Palate & (21) Other defect	69	26	73	47	44	15	44	18	62·0	45·2	9	2	25	16	24·0	24·6	39	15	35	15	52·1	41·1
,, & (11) Ear	51	7	52	10	32	5	24	4	54·3	52·8	3	4	12	6	14·5	58·8	25	2	16	3	39·8	29·4
,, & (12) Epicanthis	18	11	23	22	7	5	8	13	36·5	54·5	3	1	5	7	19·5	24·2	6	7	7	6	31·7	39·3
(11) Ear & (21) Other defect	61	14	45	13	43	4	29	9	68·0	48·1	9	2	11	7	18·8	33·3	25	5	18	6	40·5	40·7
,, & (12) Epicanthis	28	13	46	16	14	7	18	9	43·2	55·1	4	4	7	8	14·8	41·3	6	6	15	5	28·3	38·9
(12) Epicanthis & (21) Other defect	46	19	54	54	25	12	27	23	52·0	48·0	2	3	11	11	13·0	19·1	16	5	25	20	41·0	34·2

TABLE XIX. (Cases seen 1888-91).—*Proportionate distribution of Groups of cases, and certain defects in the Divisions of Schools, showing the relative condition of the Children in the Resident and Day Schools, the Social classes, and Nationalities.*

The numbers of cases are obtainable as follows: For the Defects, see Sub-Table and Table IX. on page 94; for the Groups of cases, see Sub-Table and Table X. on page 96; for all the Schools, see Defects in Nomenclature on page 72; for Groups of cases, see Catalogue on page 82.

Percentages are taken upon the numbers of children seen, as given in the List of Divisions of Schools on page 5.

N.B.—A further distribution among the Nationalities is given on Table XX.

Divisions of Schools. See List, on p. 5.	Children noted. Group 2. B.	Children noted. Group 2. G.	Eye cases. Group 3. B.	Eye cases. Group 3. G.	Eye cases. Group 3. R.	Dev. defect alone or in comb. Group 28. G.	Dev. defect alone or in comb. Group 28. R.	Nerve defect alone or in comb. Group 29. B.	Nerve defect alone or in comb. Group 29. G.	Low Nutrition alone or in comb. Group 30. B.	Low Nutrition alone or in comb. Group 30. G.	Dull alone or in comb. Group 31. B.	Dull alone or in comb. Group 31. G.	Dev. and Nerve. Group 32. B.	Dev. and Nerve. Group 32. G.	Dev., Nerve, and Low Nutrition. Group 38. B.	Dev., Nerve, and Low Nutrition. Group 38. G.	Dev. without Nerve. Group 42. B.	Dev. without Nerve. Group 42. G.	Nerve without Dev. Group 43. B.	Nerve without Dev. Group 43. G.	Small Head. Defect (3). B.	Small Head. Defect (3). G.	Growth Small. Defect (20). B.	Growth Small. Defect (20). G.
1. Poor Law Schools	22·6	17·3	3·5	2·6	15·0	12·2	15·0	15·0	9·8	4·0	2·3	8·7	7·1	9·2	6·2	1·6	1·2	5·9	5·8	6·9	3·5	0·7	1·7	0·7	1·2
2. Certified Industrial Schools	31·4	22·3	3·4	4·9	20·7	15·2	20·7	20·8	14·0	2·5	6·8	14·0	11·5	12·6	9·0	1·2	3·4	8·1	6·1	8·3	4·9	1·5	6·3	1·3	1·4
3. Homes and Orphanages	22·2		3·6	2·6	13·8	12·8	13·8	14·5	10·3	1·8	3·3	8·8	10·8	9·9	6·5	0·5	1·4	3·9	6·3	4·5	4·1	1·8	5·0	0·4	0·8
4. All Resident Schools	24·1	17·7	3·4	3·0	16·3	12·4	16·3	16·2	10·3	3·5	2·8	9·7	8·1	9·9	6·5	1·4	1·4	6·1	5·9	6·2	3·7	0·9	2·7	0·8	1·2
5. Public Elementary Day Schools	19·1	14·9	2·9	2·7	12·2	8·8	12·2	11·1	8·6	3·9	5·6	7·6	5·7	6·2	4·6	2·0	1·9	7·3	3·8	4·5	4·4	1·3	2·7	0·8	0·8
6. Day Schools, Upper Social class	21·1	16·1	3·2	2·9	14·3	8·5	14·3	12·7	10·5	4·6	4·8	8·7	5·9	7·0	4·2	1·4	1·6	6·3	4·6	5·1	5·5	1·1	2·3	0·8	0·5
7. Day Schools, Poorer Social class	18·3	14·4	2·8	2·7	12·6	9·0	12·6	11·0	7·7	3·6	4·8	7·8	5·8	6·2	4·2	1·6	1·7	6·3	4·9	4·8	3·9	1·4	2·3	0·7	0·9
8. English Day Schools	19·2	15·0	2·9	2·7	13·4	9·0	13·4	12·7	9·9	3·8	4·2	8·2	6·3	7·3	4·7	1·5	1·6	6·1	4·9	5·3	4·2	1·2	3·2	0·8	0·9
9. All the Irish Schools	34·6	19·3	4·8	4·5	22·4	9·0	22·4	25·7	11·9	4·0	5·9	14·3	9·7	16·5	4·0	2·3	2·1	3·8	4·7	9·2	4·2	1·3	3·4	0·8	0·8
10. Jewish Day School	17·8	13·9	3·3	2·2	11·2	6·9	11·2	11·2	8·4	3·0	2·4	4·6	5·6	4·4	2·9	0·8	0·6	4·0	7·0	6·7	5·5	1·1	2·3	1·1	0·8
All the Schools seen	20·8	15·6	3·1	2·7	13·4	9·7	13·4	12·7	9·9	3·8	4·2	8·2	6·3	7·3	4·7	1·5	1·6	6·1	4·9	5·3	4·2	1·2	3·2	0·8	0·9

Figures used in obtaining above percentages, and not printed elsewhere:—

Division	G2 B	G3 B	G3 R	G28 G	G28 R	G29 B	G29 G	G30 B	G30 G	G31 B	G31 G	G32 B	G32 G	G38 B	G38 G	G42 B	G42 G	G43 B	G43 G	SH B	SH G	GS B	GS G	Rickets	Gr. p. F.	Feeble-minded	Groups 6,7,8,9
6. Day Schools, Upper Social class	168	168	146	419	756	672	521	257	276	462	290	370	228	107	94	386	191	238	269	60	112	30	27	32	13	22	15
7. Day Schools, Poorer Social class	381	381	332	1154	1536	1408	996	482	541	953	732	785	515	186	211	751	630	687	505	187	480	114	119	81	81	36	35
8. English Day Schools	490	490	427	…	…	…	…	…	…	…	…	1055	675	269	284	1075	749	805	675	230	551	126	131	106	10	70	48
9. All the Irish Schools	…	…	…	…	…	…	…	…	…	…	…	280	35	39	17	100	42	156	48	13	14	15	5	13	1	6	1
10. Jewish Day School	…	…	…	…	…	…	…	…	…	…	…	62	46	12	19	53	62	95	86	15	35	15	12	7	1	4	

TABLE XX. (*Cases seen 1888-91*).—*Proportional Distribution of Groups of cases among the Nationalities, showing the Numbers of Children of various Nationalities in Divisions of Schools, and their relative condition.*

For a description of Divisions of Schools, see List on page 5; for definition of Defects, see Nomenclature on page 72; and for Groups of cases, see Catalogue on page 82. Percentages are taken upon the number of Children seen.

Divisions of Schools. See List on page	Number of Children seen		Children noted Group 2				With Defect in Development, alone or in combination. Grp. 28.				With Nerve-signs, alone or in combination. Group 29.				With Low Nutrition, alone or in combination. Group 30.				With Mental Dulness, alone or in combination. Group 31.			
	Boys	Girls	No. B.	No. G.	% B.	% G.	No. B.	No. G.	% B.	% G.	No. B.	No. G.	% B.	% G.	No. B.	No. G.	% B.	% G.	No. B.	No. G.	% B.	% G.
English Children—																						
8. Day	16,932	15,875	3252	2379	19.2	15.0	2129	1424	12.6	9.0	1660	1351	11.0	8.5	673	750	4.0	4.8	1326	915	7.8	5.8
Poor Law	5,083	3,748	1109	647	21.8	17.3	739	449	14.5	12.1	722	366	14.2	9.9	207	89	4.1	2.4	412	261	8.1	7.0
Certified Industrial	1,012	304	214	62	21.1	20.4	145	43	14.3	14.1	128	42	12.6	13.8	11	16	1.1	5.3	103	28	10.2	9.2
3. Homes and Orphanages	774	1,049	172	186	22.2	17.7	107	134	13.8	12.8	112	112	14.5	10.7	14	35	1.8	3.3	68	113	8.8	10.8
Total English Children	23,801	20,976	4747	3274	19.9	15.6	3120	2050	13.1	9.8	2822	1871	11.9	8.9	905	900	3.8	4.3	1909	1317	8.0	6.3
Irish Children—																						
Day	317	293	76	48	24.0	16.4	47	32	14.8	10.9	65	34	20.5	11.6	24	19	7.6	6.5	25	19	7.9	6.5
Poor Law	801	199	223	38	27.8	19.1	149	26	18.6	13.1	167	22	20.9	11.1	29	4	3.6	2.0	98	20	12.2	10.0
Certified Industrial	576	103	286	29	49.7	28.2	184	19	32.0	18.4	204	15	35.4	14.6	30	12	5.2	11.7	120	19	20.8	18.4
9. Total Irish Children	1,694	595	585	115	34.5	19.3	380	77	22.4	12.9	436	71	25.7	11.9	83	35	4.9	5.9	243	58	14.3	9.7
Jewish Children—																						
10. Day	1,389	1,574	247	218	17.8	13.9	116	108	8.4	6.9	155	132	11.2	8.4	42	38	3.0	2.4	64	88	4.6	5.6
Grand Total, all Children	26,884	23,143	5579	3607	20.8	15.6	3616	2235	13.4	9.7	3413	2074	12.7	9.0	1030	973	3.8	4.2	2216	1463	8.2	6.3

	Cranium defect (1).				Palate defect (13).				External Ear defect (11).				Epicanthis def't. (12).				Other defects (21).			
	No. of cases		Per cent.		No. of cases		Per cent.		No. of cases		Per cent.		No. of cases		Per cent.		No. of cases		Per cent.	
	Boys	Girls	Boys	Girls	Boys	Girls	Boys	Girls	Boys	Girls	Boys	Girls	Boys	Girls	B's.	G's.	Boys	Girls	Boys	Girls
8. English Children, Day Schools.	872	609	5.2	3.8	461	317	2.7	2.0	620	750	3.7	4.7	337	245	2.0	1.5	457	378	2.7	2.4
9. All Irish Children	190	31	11.2	5.2	80	28	4.7	4.7	117	8	6.9	1.3	37	8	2.2	1.3	132	35	7.7	5.8
10. All Jewish Children	47	51	3.4	3.2	16	19	1.1	1.2	42	22	3.0	1.4	4	11	0.3	0.7	33	31	2.4	1.9

TABLE XXI. (*Cases seen 1892-94*).—*Distribution of Cases noted, into Primary groups, arranged according to Standards.*

Showing all "Cases Noted" arranged in Primary groups of defects, according to Educational Standards. Also giving the "Numbers seen," "Numbers noted" and the percentage of the latter on the former, for each Standard respectively. Column headed "Total Number" gives the number in each Primary group.

For definition of Primary groups, see Catalogue, on page 82.

This Table was prepared by sorting all the cards according to Primary groups of defects, the numbers being entered in the right-hand column. Each group was then sorted or distributed according to Standards, and the numbers were entered in their respective columns.

Grp. Symbol	Infants B	Infants G	Standard O. B	Standard O. G	Standard I. B	Standard I. G	Standard II. B	Standard II. G	Standard III B	Standard III G	Standard IV. B	Standard IV. G	Standard V. B	Standard V. G	Standard VI. B	Standard VI. G	Standard VII B	Standard VII G	Standard Ex. VII B	Standard Ex. VII G	No Standard B	No Standard G	Total Number B	Total Number G
13. A	204	168	6	3	146	68	106	63	98	59	83	30	80	27	56	18	14	8	4	1	6	...	802	415
14. B	115	58	13	9	180	107	190	139	185	146	145	117	108	73	72	76	25	32	4	5	22	...	1059	762
15. C	35	36	4	2	28	21	11	15	13	7	8	9	3	7	5	6		6	1	1		...	103	110
16. D	83	57	12	17	70	56	43	63	51	41	32	9	18	25	14	6	6	...	1	1	1	...	381	297
17. A	51	32	9	7	87	55	84	39	56	34	45	28	40	17	19	3	12	...	3	...	9	...	415	207
18. A B	60	73	1	1	24	35	23	24	7	9	8	19	25	7	4	4	4	1		2	1	...	134	162
19. A C	103	84	12	26	82	54	56	55	58	44	35	9	12	20	13	7	7	4	1	1	2	...	394	314
20. A D	25	17	2		25	22	20	22	13	15	10	24	12	8	6	14	2	3				...	115	109
21. B C	113	59	30	26	147	127	138	88	115	88	88	56	42	25	20	1	1		1		8	...	703	487
22. B D	26	12	3	6	11	10	7	9	10	6	4	6	1	3	2	3				1	1	...	63	53
23. C D	27	23	1	19	14	14	10	13	10	10	1	9	3	3	4	2	2	3				...	69	77
24. A B D	56	39	19	4	85	57	61	40	49	36	24	18	17	9	10	1				1		...	323	224
25. A C D	33	46	9	4	21	33	15	11	7	7	3	4	1	1	2	3		1	1		2	...	91	110
26. B C D	31	11	3	4	20	23	15	11	8	13	5	4	2	2	2	2						...	89	70
27. A B C D	25	24	5	2	29	38	6	5	8	3	4	2	2	1	1	1		1				...	80	79
66. E F G	85	76	2	...	47	24	32	36	43	44	47	46	34	37	26	31	9	23	6	4	5	...	336	323
Number noted Group 2 (N)	1072	815	130	135	1016	744	817	632	727	569	542	392	390	265	255	178	83	81	23	18	57	...	5112	3829
Number seen (S)	7055	6274	354	323	4231	4191	3710	3465	3541	3434	2760	2534	2195	1797	1471	1078	508	486	144	131	228	...	26,287	23,713
Percentage of cases noted on number seen	15.1	13.0	36.7	41.7	24.0	17.7	22.0	18.2	20.5	16.5	19.6	15.4	17.7	14.7	17.3	16.5	13.7	16.6	16.0	13.7	25.0	...	19.4	16.1

TABLE XXII. (*Cases seen 1892-94*).—*Distribution of Cases noted, into Primary groups, arranged according to Ages.*

Showing all "Cases Noted" arranged in Primary groups of Defects, according to Ages. Also giving the numbers noted, for each Age-group respectively.

This Table was prepared by sorting cards in Primary groups, sub-dividing them according to Age-groups, and entering the numbers in the respective columns.

For definition of Primary groups, see Catalogue on page 82.

The numbers of the Primary groups in right-hand column agree with those in the Catalogue.

Primary Groups of Defect.		3 and under. B.	G.	4. B.	G.	5. B.	G.	6. B.	G.	7. B.	G.	8. B.	G.	9 10. B.	G.	11–12. B.	G.	13. B.	G.	14 and over. B.	G.	Total Numbers. B.	G.
Grp.	Symbol.																						
13.	A	10	22	30	27	69	70	100	44	100	60	78	50	190	93	183	60	31	13	11	6	802	445
14.	B	7	1	13	3	25	19	70	35	124	70	155	104	310	239	284	230	48	41	23	20	1039	762
15.	C	1	1	5	3	14	12	19	16	18	17	12	13	23	20	12	18	3	8	1	1	108	110
16.	D	3	2	12	9	28	17	31	27	42	30	29	38	77	73	77	75	22	19	10	7	311	297
17.	A B	2	…	5	4	18	11	31	23	66	26	67	40	104	55	97	40	26	7	9	1	415	207
18.	A C	9	10	11	17	14	29	24	17	13	26	23	24	23	26	12	11	3	2	2	…	134	162
19.	A D	6	3	21	13	35	32	40	32	36	37	52	49	92	77	86	56	19	10	7	0	394	314
20.	B C	1	1	2	1	8	4	13	13	19	16	19	13	25	30	26	23	5	6	…	5	115	109
21.	B D	2	1	13	0	32	13	55	33	68	59	106	63	202	145	172	126	41	38	13	3	708	487
22.	C D	…	4	3	4	10	4	10	10	11	11	5	2	14	14	8	12	2	2	…	…	63	53
23.	A B C	…	…	7	…	8	5	10	12	10	28	10	11	15	17	7	11	2	12	…	…	69	77
24.	A B D	2	3	11	3	13	9	14	16	46	25	59	31	86	66	74	50	16	2	4	6	328	224
25.	A C D	1	…	4	3	10	15	15	24	19	14	13	9	20	19	7	11	…	1	1	…	91	110
26.	B C D	3	3	3	…	6	4	16	2	12	17	14	14	21	23	10	10	2	2	2	…	89	70
27.	A B C D	…	…	4	3	6	7	9	12	11	17	17	9	18	16	10	7	2	1	…	1	80	79
56.	E F G	9	8	12	8	21	30	41	23	29	22	24	23	90	82	81	84	20	26	9	12	336	323
2.	Number noted N …	67	59	156	113	320	281	492	339	614	466	683	493	1310	995	1146	824	242	190	92	69	5112	3829

TABLE XXIII. (*Cases seen 1892-94*).—*Showing the proportion of the Primary groups of cases to the Compound groups respectively.*

The Compound groups are those containing all cases presenting the defects or combination of defects indicated alone or in combination. They are obtained by addition of all the Primary groups containing the defect, or combination of defects indicated (see Catalogue on page 82, Groups 28 to 41, and symbols in margin).

For definition and numbers in each Primary group, see Catalogue, page 82. As already said, it is by addition of numbers in Primary groups that the Compound groups are obtained; the figures are thus afforded by which the percentage distribution of the Primary groups upon the Compound groups are prepared.

Grp. Symbol.	Group 13 A B.	G.	Group 14 B B.	G.	Group 15 C B.	G.	Group 16 D B.	G.	Group 17 A B B.	G.	Group 18 A C B.	G.	Group 19 A D B.	G.	Group 20 B C B.	G.	Group 21 B D B.	G.	Group 22 C D B.	G.	Group 23 A B C B.	G.	Group 24 A B D B.	G.	Group 25 A C D B.	G.	Group 26 B C D B.	G.	Group 27 A B C D B.	G.	Group 56 E, F, or G. B.	G.
Number seen S / 2. noted N	340 187	275 116	40 207	32 199	4 21	3 29	12 65	12 78	16 81	9 54	5 26	7 42	15 77	13 82	4 22	5 28	27 138	20 127	2 12	2 14	3 14	3 20	12 63	9 59	8 18	5 29	3 18	3 18	3 16	3 21	19·2 100·0	16·1 100·0
28. A +	34·7	27·5	...	...	...	...	...	...	18·0	12·8	8·8	10·0	17·1	19·4	...	...	...	...	...	...	3·0	4·8	14·0	13·8	3·9	6·8	...	...	3·5	4·9	100·0	100·0
29. B + +	...	...	37·1	37·8	...	...	...	...	14·6	10·3	...	...	...	...	4·0	...	24·7	24·2	...	...	2·4	3·8	11·3	11·1	...	...	3·1	3·5	2·8	3·9	100·0	100·0
30. C + +	...	...	...	...	14·4	14·3	...	...	...	...	17·9	21·0	...	...	15·4	...	...	...	8·4	6·9	9·2	10·0	...	...	12·1	14·3	9·1	9·1	10·7	10·2	100·0	100·0
31. D +	...	...	...	...	...	...	16·0	18·2	...	...	...	...	19·0	19·2	...	...	33·9	29·8	3·0	3·3	...	...	15·6	13·7	4·4	6·7	4·3	4·3	3·8	4·8	100·0	100·0
32. A B	...	...	...	...	...	...	...	...	46·8	35·3	...	...	...	...	...	...	...	...	...	...	7·7	13·1	36·4	38·1	...	...	...	...	9·1	13·5	100·0	100·0
33. A C	...	...	...	...	...	...	...	...	...	...	35·8	37·8	...	...	...	...	...	...	...	...	18·5	18·0	...	...	24·8	25·7	...	...	21·4	18·5	100·0	100·0
34. A D	...	...	...	...	...	...	...	...	...	...	...	...	44·4	43·2	...	...	...	...	...	...	...	...	36·4	30·8	10·2	15·1	...	...	9·0	10·9	100·0	100·0
35. B C	...	...	...	...	...	...	...	...	...	...	...	...	...	...	92·6	32·6	...	...	...	...	19·5	23·0	...	...	...	...	22·7	20·8	22·7	23·6	100·0	100·0
36. B D	...	...	...	...	...	...	...	...	...	...	...	...	...	...	...	...	58·8	55·6	...	...	...	...	27·0	26·1	...	...	7·5	8·1	6·7	9·2	100·0	100·0
37. C D	...	...	...	...	...	...	...	...	...	...	...	...	...	...	...	...	...	...	19·5	17·0	46·3	49·4	...	...	28·2	35·3	27·5	22·4	24·8	25·3	100·0	100·0
38. A B C	...	...	...	...	...	...	...	...	...	...	...	...	...	...	...	...	...	...	...	...	...	...	...	...	...	...	...	...	53·7	50·6	100·0	100·0
39. A B D	...	...	...	...	...	...	...	...	...	...	...	...	...	...	...	...	...	...	...	...	...	...	80·1	73·9	...	...	...	...	19·9	26·1	100·0	100·0
40. A C D	...	...	...	...	...	...	...	...	...	...	...	...	...	...	...	...	...	...	...	...	...	...	...	...	64·	58·2	...	...	46·8	41·8	100·0	100·0
41. B C D	...	...	...	...	...	...	...	...	...	...	...	...	...	...	...	...	...	...	...	...	...	...	...	...	...	...	52·7	47·0	47·3	53·0	100·0	100·0
42. (A+)−(AB+)	56·5	43·2	...	...	...	...	...	...	...	...	...	...	...	...	...	...	...	...	...	...	...	...	...	...	...	...	...	...	...	...	100·0	100·0
45. (B+)−(AB+)	...	...	53·9	53·4	...	...	...	...	...	...	9·4	15·7	27·7	30·5	6·8	7·6	85·7	34·1	12·5	11·5	...	...	...	...	6·4	10·6	52·7	47·0	46·8	53·0	100·0	100·0
54. See Catalogue...	...	...	...	...	21·5	23·9	66·0	64·6	...	...	...	...	...	...	...	...	...	...	...	...	...	...	...	...	...	...	4·5	4·9	47·3	...	100·0	100·0

TABLE XXIV. (*Cases seen 1892-94*).—*Co-relation of Compound groups with other defects, alone or in combination.*

The method of obtaining the Compound groups is explained in note appended to Table XXIII, which also affords explanation of the method of obtaining the number of cases with any selected defect or group of defects contained in a Compound group.

The percentages are the numbers in the group denoted by the symbols in the first column and head-line together, as compared with the numbers in the group denoted by the symbol in the first column alone.

For definition of groups and their numbers, see Catalogue, page 82.

Percentages Grp. Symbol	A+		B+		C+		D+		AB+		AC+		AD+		BC+		BD+		CD+		ABC+		ABD+		ACD+		BCD+		ABCD		E F or G		G.	
	B.	G.	B.	G.	B.	G.	B.	G.	B.	G.	B.	G.	B.	G.	B.	G.	B.	G.	B.	G.	B.	G.	B.	G.	B.	G.	B.	G.	B.	G.	B.	G.	B.	G.
Number seen S	8·8	6·8	10·9	8·5	2·8	3·2	7·9	6·9	3·4	2·5	1·4	1·8	3·4	3·1	1·8	1·4	4·5	3·6	1·2	1·3	0·6	0·7	1·5	1·3	0·6	0·8	0·6	0·6	0·3	0·3			1·3	1·4
„ noted N	45·1	42·2	55·8	52·6	14·7	20·1	40·6	42·6	17·3	15·3	7·3	11·2	17·3	18·9	6·9	8·7	23·4	22·5	6·3	8·1	2·9	4·1	7·0	7·9	8·8	4·9	3·8	3·9	1·6	2·1			6·8	8·4
28. A+			38·1	36·3	16·2	26·5	38·4	44·9							6·4	9·6	17·4	18·7	7·4	11·6							3·5	4·9						
29. B+	31·0	29·1			12·3	16·6	41·8	42·6					14·1	15·0					5·9	7·4														
30. C+	40·9	55·6	47·1	43·5			43·1	40·5	19·9	20·3	5·2	7·7			8·1	9·1	22·6	19·3			3·9	4·8			2·8	3·9								
31. D+	42·8	44·5	57·6	52·6	15·5	19·1			19·4	18·5	8·2	11·6	22·8	24·5					9·0	13·5			10·7	10·3										
32. AB+							45·5	51·6																										
33. AC+					16·8	26·6																												
34. AD+			39·8	36·4			45·7	44·2							9·0	10·9	21·4	18·4																
35. BC+	42·2	46·6	45·4	41·7	19·2	26·0					6·7	9·2																						
36. BD+	33·7	35·2			14·3	17·3	47·0	44·5					22·6	23·6																				
37. CD+	52·9	60·6	52·3	47·7					24·8	25·3																								
38. ABC+					19·8	26·1																												
39. ABD+							53·7	50·6																										
40. ACD+			46·8	41·8																														
41. BCD+	47·3	53·0																																
42. (A+)−(AB+)					15·8	26·4	34·1	41·1											6·4	10·6														
45. (B+)−(AB+)					10·4	12·5	42·2	39·0											4·5	4·9														
54. See Catalogue					34·0	35·4	78·5	76·1											12·6	11·5														

Some of these Co-relations of Cases seen 1888-91 are given in Tables XVI. and XVII.

TABLE XXV. (Cases seen 1892-94).—Showing the relation of Age-groups to Standard-groups among the cases noted.

The numbers noted in the Standards respectively are entered from Table XXI., and the numbers in Age-groups from Table XXII. Numbers noted in the Standards are distributed in second and third columns into those of proper age for the Standard, and those over age respectively. Numbers of children in second column arranged according to Age-groups are distributed in first and fifth columns; those of proper age and those presumably above age for their Standards respectively.

Percentages in fourth column are those of figures in third column taken on numbers in first column.

Percentages in last column are those of numbers in fifth column on numbers in second column.

Cases noted. The recognized age of children in each Standard is shown below.*	Number noted in Standard.		Number of Standard age.		Number presumably over Standard age.		Per cent. of children noted in Standard presumably over age.		Numbers in Age-groups presumably in too low a Standard.		Per cent. in age-groups presumably in too low a Standard.	
	Boys.	Girls.	Boys.	Girls.	Boys.	Girls.	Boys.	Girls.	Boys.	Girls.	Boys.	Girls.
Infants	1,072	815	...	...	539	362	50·3	44·4	...	...	...	...
Children aged 5 and under	...	...	533	453	...	...	...	...	...	...	...	...
Standard I.	1,016	744	...	...	524	405	51·6	54·4	...	...	...	...
Children aged 6	...	...	492	339	...	...	...	...	...	...	...	...
Standard II.	817	632	...	...	203	166	24·8	26·3	...	...	...	...
Children aged 7	...	...	614	466	...	...	...	...	...	...	...	...
Standard III.	727	569	...	...	44	76	6·1	13·4	...	...	...	...
Children aged 8	...	...	683	493	...	...	...	...	...	...	...	...
Standards IV., V.	932	657	...	...	...	...	...	...	...	...	...	...
Children aged 9, 10	...	...	1,310	995	...	...	...	...	378	338	28·8	33·9
Standards VI., VII.	338	259	...	...	...	...	...	...	...	...	...	...
Children aged 11, 12	...	...	1,146	824	...	...	...	...	808	565	70·6	68·6
Standard ex VII.	23	18	...	...	...	...	...	...	...	...	...	...
Children aged 13 and over	...	...	334	259	...	...	...	...	124	106	37·1	40·9
Standard O	130	135	...	...	...	...	...	...	...	...	...	...
Not in Standards	57	...	...	...	...	...	...	...	...	...	...	...
TOTALS	5,112	3,829	5,112	3,829	1,310	1,009	25·6	26·3	1,310	1,009	25·6	26·3

* The Committee have been informed that the age officially recognised for children in the several Standards is as stated in the above Table.

TABLE XXVI. (Cases seen 1892-94).—Distribution of Cases in Divisions of Schools arranged according to Nationalities, Social Classes, London Board Schools, &c., under Standards giving number of children seen and numbers noted.

For description of Divisions of Schools, see List on page 7.

Divisions of Schools.		Infants B.	Infants G.	Std 0 B.	Std 0 G.	Std I B.	Std I G.	Std II B.	Std II G.	Std III B.	Std III G.	Std IV B.	Std IV G.	Std V B.	Std V G.	Std VI B.	Std VI G.	Std VII B.	Std VII G.	Ex. VII B.	Ex. VII G.	No Std B.	No Std G.	Total seen B.	Total seen G.	Total noted B.	Total noted G.
Upper Social Class — Divisions 1, 6, 7.	No. seen	1635	1413	66	55	904	888	882	757	876	822	747	710	659	577	443	432	284	288	111	120	228	…	6,835	6,062	…	…
	No. noted	260	164	26	10	152	145	183	121	183	183	151	121	109	93	77	73	41	44	16	16	57	…	…	…	1251	969
Average Social Class — Divisions 2, 5, 8, 11.	No. seen	2292	1900	102	110	1355	1384	1278	1270	1101	1168	940	874	705	616	460	318	182	145	17	6	…	…	8,432	7,791	…	…
	No. noted	329	265	29	47	348	225	301	234	228	189	184	141	122	87	58	52	26	31	6	2	…	…	…	…	1630	1276
Poorer Social Class — Divisions 3, 4, 9, 10, 12.	No. seen	3128	2961	186	138	1972	1919	1550	1438	1561	1444	1073	950	831	604	568	348	132	53	16	5	…	…	11,020	9,860	…	…
	No. noted	483	386	75	78	516	371	231	277	216	197	207	130	166	86	120	53	15	6	2	…	…	…	…	…	2231	1584
English Children — Divisions 1, 2, 3, 6, 7, 8, 9.	No. seen	5516	4732	339	264	3311	3244	2946	2697	2707	2609	2206	2001	1689	1395	1107	828	505	404	128	112	228	…	20,632	18,256	…	…
	No. noted	781	596	125	110	773	584	642	525	547	440	433	308	270	202	163	125	63	67	18	15	57	…	…	…	3872	2372
Scotch Children — Division 6.	No. seen	173	165	…	…	87	102	150	103	107	125	117	103	88	98	50	59	20	33	11	19	…	…	803	807	…	…
	No. noted	48	32	…	…	11	13	36	13	16	24	19	13	12	16	10	10	4	8	4	3	…	…	…	…	163	128
Irish Children — Division 10.	No. seen	629	613	15	…	451	438	336	330	287	232	198	187	158	107	78	49	19	6	…	…	…	…	2,171	1,952	…	…
	No. noted	121	79	5	…	143	84	84	55	70	39	49	37	38	19	22	19	3	…	6	1	…	…	…	…	535	324
Jew Children — Divisions 4, 11, 12.	No. seen	737	764	…	59	382	407	278	345	440	458	239	243	260	239	236	142	54	43	6	1	…	…	2,631	2,668	…	…
	No. noted	122	108	…	25	89	62	55	39	95	66	41	38	71	29	66	33	12	5	1	…	…	…	…	…	542	405
London Board Schools — Divisions 1, 2, 3, 4.	No. seen	5224	4530	339	323	3071	2888	2626	2418	2408	2348	1959	1765	1464	1238	983	766	425	361	124	107	…	…	18,623	16,738	…	…
	No. noted	739	569	125	135	712	537	568	471	449	382	368	263	225	168	142	110	65	59	18	15	…	…	…	…	3401	2709
Schools of special difficulty. See page 61.	No. seen	353	324	13	13	170	190	152	143	124	146	96	101	63	57	27	22	4	…	…	…	…	…	1,022	996	…	…
	No. noted	45	47	7	6	70	31	45	41	61	61	22	20	10	13	8	12	…	14	…	…	…	…	…	…	227	163
Schools with few children in relation to accommodation. See page 64.	No. seen	142	120	13	13	99	75	97	83	…	61	32	69	49	49	32	12	14	14	…	…	…	…	666	451	…	…
	No. noted	84	25	7	6	81	25	27	21	15	10	…	17	11	11	6	1	1	1	…	…	…	…	…	…	149	96

H

TABLE XXVII. (*Cases seen 1892-94*).—*Distribution of cases in Divisions of Schools, arranged according to Nationalities, Social Classes, London Board Schools, &c., under Primary Groups of Defects.*

For description of Divisions of Schools, see List on page 7.   For definition of Primary Group of Defects, see Catalogue on page 82.

Divisions of Schools.	Group 13 A		Group 14 B		Group 15 C		Group 16 D		Group 17 AB		Group 18 AC		Group 19 AD		Group 20 BC		Group 21 BD		Group 22 CD		Group 23 ABC		Group 24 ABD		Group 25 ACD		Group 26 BCD		Group 27 ABCD		Group 55 EFG		Total Noted.	
	R.	G.	R.	G.	R.	G.	R.	G.	R.	G.	R.	G.	R.	G.	R.	G.	R.	G.	R.	G.	R.	G.	R.	G.	R.	G.	R.	G.	R.	G.	R.	G.	R.	G.
Upper Social Class Divisions 1, 6, 7.	220	114	289	243	24	27	70	57	102	54	27	30	95	74	28	31	160	114	8	13	11	17	64	39	20	24	13	11	14	12	103	109	1221	969
Average Social Class Divisions 2, 5, 8, 11.	229	135	334	247	35	39	114	107	138	73	43	51	120	110	36	27	296	173	91	18	22	27	112	78	28	38	33	31	27	20	92	102	1630	1276
Poorer Social Class Divisions 3, 4, 9, 10, 12.	353	196	436	272	49	44	147	133	175	80	64	81	179	130	51	51	307	200	24	22	33	33	147	107	43	48	43	28	39	47	141	112	2231	1584
English Children Divisions 1, 2, 3, 5, 7, 8, 9.	619	360	766	583	78	88	249	235	297	150	105	126	298	241	83	79	546	389	55	43	51	60	237	164	90	77	68	59	65	59	268	246	3872	2972
Scotch Children Division 6.	16	6	39	38	8	8	10	5	14	6	2	3	15	7	6	4	29	21	1	…	…	1	13	7	5	1	1	4	3	2	9	11	163	128
Irish Children Division 10.	91	40	123	56	14	6	23	13	60	24	12	16	29	27	12	15	66	33	3	4	6	6	37	25	6	10	10	3	8	12	32	31	535	324
Jewish Children Divisions 4, 11, 12.	73	39	131	85	12	8	39	44	44	27	15	17	52	39	14	11	62	44	4	6	12	7	36	28	10	7	10	4	4	6	27	35	542	405
London Board Schools. Divisions 1, 2, 3, 4.	552	322	612	516	71	85	246	231	219	132	98	116	287	230	65	77	484	339	45	39	45	56	222	151	68	85	64	52	59	57	234	221	3401	2709
Schools of special difficulty. See p. 64.	33	20	27	24	6	6	21	22	22	2	7	9	23	13	4	5	30	20	2	4	2	2	19	16	9	8	6	4	6	3	13	9	227	169
Schools with few children in relation to their accommodation. See p. 64.	13	6	32	9	1	1	17	11	12	9	2	6	11	12	4	1	17	15	1	1	3	3	16	1	6	5	4	3	5	5	8	8	149	96
All Standard O. See p. 63	5	3	13	3	4	4	12	12	17	7	1	1	12	26	2	2	30	26	3	6	2	1	19	19	9	4	3	4	5	5	2	2	130	135
All Standard Ex. VII. See p. 63	4	1	4	5	1	1	2	1	3	3	…	…	1	2	1	1	1	1	…	1	…	1	1	1	…	1	3	4	5	6	6	4	23	18
Imbeciles. Group 7	…	…	…	…	…	…	…	…	…	…	…	…	…	…	…	…	…	…	…	…	…	…	…	…	…	…	…	…	…	…	…	…	3	2
Mentally feeble. Group 8	…	…	…	…	…	…	4	3	…	…	…	…	5	4	…	…	17	17	…	…	…	…	…	16	…	2	3	…	4	8	…	…	49	52
Mentally exceptional. Group 9	…	…	…	…	…	…	…	…	…	…	…	…	…	…	…	…	1	…	…	…	…	…	…	1	…	…	1	1	…	…	…	…	4	4
Epileptic. Group 10	…	5	3	…	…	…	1	…	…	…	…	…	1	1	…	…	10	…	…	…	1	3	1	…	1	1	1	…	1	…	3	6	21	35
Crippled. Group 11	2	…	14	15	4	2	9	4	4	1	2	1	1	1	…	…	11	4	…	…	2	2	2	…	2	1	1	1	2	1	26	10	75	60

TABLE XXVIII. (Cases seen 1892-94).—*Proportional distribution of groups of cases in Divisions of Schools arranged according to Nationalities, Social Classes, London Board Schools, &c., showing the numbers of children and their relative condition.*

For description of Divisions of Schools, see List on page 7. For definition of Groups of Defects, see Catalogue on page 82. Percentages are taken upon the number of children seen.

Divisions of Schools.	Numbers seen.		Development Defect, alone or in combination. Group 28.				Nerve-signs, alone or in combination. Group 29.				Low Nutrition alone or in combination. Group 30.				Mental Dulness, alone or in combination. Group 31.				Development Defect and Nerve-signs, alone or in combination. Group 32.				Development Defect, Nerve-signs and Low Nutrition, alone or in combination. Group 34.				Development Defect without Nerve-signs, alone or in combination. Group 42.				Nerve-signs without Development Defect, alone or in combination. Group 45.			
			No. of cases.		Per cent.		No. of cases.		Per cent.		No. of cases.		Per cent.		No. of cases.		Per cent.		No. of cases.		Per cent.		No. of cases.		Per cent.		No. of cases.		Per cent.		No. of cases.		Per cent.	
	B.	G.	B.	G.	B.	G.	B.	G.	B.	G.	B.	G.	B.	G.	B.	G.	B.	G.	B.	G.	B.	G.	B.	G.	B.	G.	B.	G.	B.	G.	B.	G.	B.	G.
Upper Social Class Divisions 1, 6, 7.	6,885	6,662	556	364	8·1	6·0	684	521	10·0	8·6	118	165	2·2	2·7	444	344	6·5	5·7	194	122	2·8	2·0	28	29	0·4	0·5	362	242	5·3	4·0	490	399	7·1	6·6
Average Social Class Divisions 2, 5, 8, 11.	8,132	7,791	719	532	8·5	6·8	938	676	11·1	8·7	255	251	3·0	3·2	701	575	8·3	7·4	299	198	3·5	2·5	49	47	0·6	0·6	420	334	5·0	4·3	639	478	7·6	6·1
Poorer Social Class Divisions 3, 4, 9, 10, 12.	11,020	9,660	1033	722	9·4	7·3	1231	818	11·2	8·3	346	354	3·1	3·6	929	715	8·4	7·3	394	267	3·6	2·7	72	80	0·7	0·8	639	455	5·8	4·6	837	551	7·6	5·6
English Children Divisions 1, 2, 3, 5, 7, 8, 9.	20,692	18,286	1749	1250	8·4	6·8	2113	1543	10·2	8·4	682	604	2·8	3·4	1605	1280	8·0	7·0	650	433	3·1	2·4	116	119	0·6	0·6	1099	817	5·3	4·5	1463	1110	7·1	6·1
Scotch Children Division 6.	803	807	64	37	8·0	4·6	105	83	13·1	10·3	18	27	2·2	3·3	73	51	9·1	6·3	30	16	3·7	2·0	3	3	0·4	0·4	34	21	4·2	2·6	75	67	9·3	8·3
Irish Children Division 10.	2,171	1,952	252	163	11·6	8·4	322	177	14·8	9·0	71	75	3·3	3·9	182	127	8·4	6·5	111	70	5·1	3·6	14	21	0·6	0·1	141	93	6·5	4·8	211	107	9·7	5·5
Jew Children Divisions 4, 11, 12.	2,631	2,668	243	168	9·2	6·3	313	212	11·9	7·9	78	64	3·0	2·4	214	176	8·1	6·6	96	68	3·6	2·5	16	13	0·6	0·5	147	100	5·6	3·7	217	144	8·2	5·4
London Board Schools Divisions 1, 2, 3, 4.	18,623	16,738	1580	1149	8·5	6·9	1860	1380	10·0	8·2	515	567	2·8	3·4	1475	1184	7·9	7·1	575	396	3·1	2·4	104	113	0·6	0·7	1005	753	5·4	4·5	1225	984	6·6	5·9
Schools of special difficulty. See page 64.	1,022	996	121	75	11·8	7·5	106	86	10·4	8·6	39	43	3·8	4·3	116	98	11·3	9·8	49	25	4·8	2·5	8	7	0·8	0·7	72	50	7·0	5·0	67	61	6·5	6·1
Schools with few children in relation to accommodation. See page 64.	566	451	65	40	11·5	8·9	90	52	15·9	11·5	23	25	4·1	5·5	77	53	13·6	11·7	33	11	5·8	2·4	5	8	0·9	1·8	32	29	5·6	6·4	57	42	10·0	9·3
Schools not under the London Board	7,664	6,975	728	469	9·5	6·7	1054	635	13·7	9·1	134	203	3·1	2·9	599	450	7·8	6·5	312	191	4·1	2·7	45	43	0·6	0·6	416	278	5·4	4·0	471	444	6·1	6·4

TABLE XXIX. (Cases seen 1892-94).—*Exceptional Children distributed into Standards under defects and under Primary groups, showing also total number of Children who appear to require special care and training.*

The analysis was prepared by sorting the cards of these cases. Group 12 includes the Exceptional Children, contained in group 5, and in Group 27. The numbers in Group 27 (Table XXI.), are therefore added to the Exceptional Children, deduction being made of the numbers of this group already included in line G with A B C D of this Table.

Nomenclature of Defects page 72. Catalogue Groups of Cases page 82.	Infants R.	Infants G.	Standard 0 R.	Standard 0 G.	Standard I. R.	Standard I. G.	Standard II. R.	Standard II. G.	Standard III. R.	Standard III. G.	Standard IV. R.	Standard IV. G.	Standard V. R.	Standard V. G.	Standard VI. R.	Standard VI. G.	Standard VII. R.	Standard VII. G.	Total Number R.	Total Number G.
Group 5 ... "Exceptional Children"	46	35	5	14	37	43	15	21	13	14	19	7	3	8	10	5	5	1	153	148
" 6 ... Idiots	2	…	…	…	1	1	…	1	…	…	…	…	…	…	…	…	…	…	3	2
" 7 ... Imbeciles	20	12	2	10	19	21	5	5	1	3	2	1	…	…	…	…	…	…	49	52
" 8 ... Children feebly gifted mentally	1	…	…	…	…	…	1	…	1	…	1	1	…	…	…	…	…	…	4	1
" 9 ... Children mentally exceptional	11	8	2	4	1	4	…	9	3	4	3	1	…	3	1	1	1	1	21	35
" 10 ... Epileptics	13	12	2	3	14	17	9	9	8	7	14	5	…	3	8	4	4	…	75	60
" 11 ... Crippled	1	1	…	…	2	…	…	…	1	…	…	…	…	…	…	…	…	…	4	1
Defect (81). Dumb	1	…	…	…	…	…	…	…	…	…	…	…	…	…	…	…	…	…	…	…
" (100). Blind, or nearly so	…	3	…	…	…	1	…	…	…	…	…	…	…	…	1	…	…	…	1	3
" (101). Chorea	…	…	…	…	…	…	…	…	…	…	…	…	…	…	…	…	…	…	…	1
" (103). Heart disease	…	…	…	…	1	…	…	…	…	…	…	…	…	…	…	…	…	…	1	…
Distribution of the "Exceptional Children" into Primary groups of Defect under Standards.																				
Symbols. G with A	1	…	…	…	…	3	…	…	…	4	1	…	…	1	1	…	1	…	4	5
G " B	3	4	1	1	2	4	3	6	1	4	3	…	…	…	1	3	2	…	14	20
G " C	1	…	1	1	1	1	…	…	1	4	…	1	…	…	…	1	…	1	4	3
G " D	6	3	1	…	3	5	3	2	2	…	6	…	…	4	1	…	1	…	22	18
G " A B	1	…	…	…	…	2	1	…	…	…	1	…	…	…	1	…	…	…	6	4
G " A C	1	…	…	…	4	…	…	…	…	…	…	…	…	…	1	…	…	…	3	…
G " A D	4	…	…	…	2	…	2	6	…	…	…	…	…	…	2	…	…	…	11	5
G " B C	…	…	…	…	…	…	…	…	…	…	…	…	1	…	…	…	…	…	…	…
G " B D	9	8	…	…	12	10	4	2	2	3	1	3	…	…	…	…	…	…	26	35
G " C D	…	…	…	…	…	1	…	1	…	…	…	…	…	…	…	…	…	…	1	1
G " A B C	…	2	…	…	…	…	…	…	2	1	…	…	…	…	…	…	…	…	…	…
G " A B D	9	5	…	3	6	7	1	4	1	…	…	…	…	…	…	…	…	…	23	19
G " A C D	…	…	…	…	1	3	…	…	…	…	…	…	…	…	…	…	…	…	1	4
G " A B C D	1	5	…	…	2	4	1	…	…	1	…	3	…	…	…	…	…	…	2	5
G " E F or G	6	7	…	…	6	…	…	…	4	…	7	…	2	1	3	1	1	…	30	17
TOTAL	46	35	5	14	37	43	15	21	13	14	19	7	3	8	10	5	5	1	153	148
To the line above may be added the cases in Group 27 not already included as "Exceptional Children"	21	19	5	5	27	34	5	5	8	3	4	2	2	1	1	1	1	…	73	70
Group 12, Children requiring special care	67	54	10	19	64	77	20	26	21	17	23	9	5	9	11	6	6	1	226	218

TABLE XXX. (*Cases seen 1892-94*).—*Exceptional children, distributed into Age-groups under defects and under Primary groups, showing also total number of children who appear to require special care and training.*

This Table is arranged on the same general plan as the last.

Nomenclature of Defects. Page 72. Catalogue or Group of Cases. Page 82.	3 and under. B.	G.	4. B.	G.	5. B.	G.	6. B.	G.	7. B.	G.	8. B.	G.	9-10. B.	G.	11-12. B.	G.	13. B.	G.	14 and over. B.	G.	Total Numbers. B.	G.
Group 5 ... "Exceptional Children"	2		6	6	11	5	12	19	14	20	17	16	33	30	33	31	18	10	7	11	153	148
,, 6 ... Idiots							1				1		1	1						1	3	2
,, 7 ... Imbeciles	1		3	3	4	1	4	6	6	9	2	10	9	9	8	2	4	1	2	4	49	52
,, 8 ... Children feebly gifted mentally						3	1	6	4	6		5	3	10	2	2	4	1	6	4	21	35
,, 9 ... Children mentally exceptional	1		1	3	5	1	3	6	4	9	6	1	18	13	20	17	11	6	6	4	75	60
,, 10 ... Epileptic			2				2	1		1	1		2		2			4		4	4	1
,, 11 ... Crippled	1		2	1			1	1		1		1										
Defect (61). Dumb																						3
,, (100). Blind, or nearly so										1		1						1				1
,, (101). Chorea	1									1												
,, (103). Heart disease			1	1	1	2	1		4		4		6	3	1		6		3		1	

Distribution of the "Exceptional Children" into Primary groups of Defect, under Age-groups.

Symbols.	3 and under. B.	G.	4. B.	G.	5. B.	G.	6. B.	G.	7. B.	G.	8. B.	G.	9-10. B.	G.	11-12. B.	G.	13. B.	G.	14 and over. B.	G.	Total Numbers. B.	G.
G with A			1	1	1		1	2	1	4	1	2	4	1	4	1	1			1	4	5
G ,, B					1		1	2	2		1		9	5	7	6	1	2	1	1	14	20
G ,, C						1			1	1			2		3	1	3	2	1		4	3
G ,, D	1		2	1	2	3	3	2	2	3		4	3	3	3	3	1	1	1	2	22	18
G ,, A B			1		1		1		1	1	1	1	1	1			2				6	4
G ,, A C													1	1		1					8	
G ,, A D														1	2						11	5
G ,, B C																						1
G ,, B D			2		2	1	3	4	2	6	8	2	5	9	3		2	3	4		26	35
G ,, C D																						1
G ,, A B C			1			1	1	1	2	3	3	3			4	4	2			4	23	2
G ,, A B D				1		1				1			1	1				1				19
G ,, A C D																					1	1
G ,, B C D	1		1	2	1	1	1	4	2	1		2		4	2			1		1	2	4
G ,, A B C D				1	1		1	2	4	1	4	2	1	3	7	1	2		7		7	5
G ,, E F or G	1		1	1	1	2	1	2					6	3	7	5	6	6		3	30	17
TOTAL	2		6	6	11	5	12	19	14	20	17	16	33	30	33	31	18	10	7	11	153	148
To the line above may be added the cases in Group 27 not already included as "Exceptional Children"	2	3	4	5	5	7	9	8	9	16	17	7	17	16	8	6	2	1		1	73	70
Group 12, Children requiring special care ...	4	3	10	11	16	12	21	27	23	36	34	23	50	46	41	37	20	20	7	12	226	218

www.ingramcontent.com/pod-product-compliance
Lightning Source LLC
Chambersburg PA
CBHW021821190326
41518CB00007B/687